Jonas Oliveira
Umberto Fulco
Eudenilson Albuquerque

Molecular Modeling

Jonas Oliveira
Umberto Fulco
Eudenilson Albuquerque

Molecular Modeling

Deciphering the Structural Stability of Human Collagen

ScienciaScripts

Imprint

Any brand names and product names mentioned in this book are subject to trademark, brand or patent protection and are trademarks or registered trademarks of their respective holders. The use of brand names, product names, common names, trade names, product descriptions etc. even without a particular marking in this work is in no way to be construed to mean that such names may be regarded as unrestricted in respect of trademark and brand protection legislation and could thus be used by anyone.

Cover image: www.ingimage.com

This book is a translation from the original published under ISBN 978-613-9-63002-8.

Publisher:
Sciencia Scripts
is a trademark of
Dodo Books Indian Ocean Ltd. and OmniScriptum S.R.L publishing group

120 High Road, East Finchley, London, N2 9ED, United Kingdom
Str. Armeneasca 28/1, office 1, Chisinau MD-2012, Republic of Moldova, Europe
Printed at: see last page
ISBN: 978-620-7-75911-8

CONTENTS

1   GENERAL INTRODUCTION

A vast amount of crystallographic data of biomolecular systems and their computational analysis are evaluated daily by researchers in research centers all over the world. This information is generated from computer simulations developed on biological structures adequately represented by theoretical and/or three-dimensional models. Traditionally, biological models are created to demonstrate how a particular system or subsystem actually works in nature through the science called *Molecular Modeling* (MM).

Molecular modeling makes it possible to build and manipulate "models" - simplified representations of real physical objects and phenomena - with the aim of gaining a deeper understanding of the entities they represent (LEACH, 2001). Thus, molecular modeling consists of computational techniques used to represent and/or realistically manipulate molecular structures in order to calculate the associated physical and chemical properties. The mathematical instrument used is theoretical chemistry, and graphical computing is the tool used to handle these models (SILVA, 2002; SANTOS, 2001).

The application of molecular modeling techniques is a complex process that begins with the construction of a reliable model; goes through the determination of the computational parameters appropriate to its objectives and the computational cost to be borne; as well as the process of screening the data obtained; and finally, formulating conclusions that must be based on consistent chemical and physical concepts, as well as reliable pre-existing experimental data.

It is therefore concluded that for a molecular modeling program to be able to adequately represent the properties of a specific molecule, it intuitively makes use of the principle of transferability, related to the prediction of properties of real systems from those obtained in representative models. To do this, it needs to correctly recognize the entire atomic structure (bonds and hybridizations) of the model studied, in order to compare it specifically with parameters from molecules used in its programming that have similar structural and electronic characteristics (RODRIGUES, 2001.).

The most widespread molecular modeling techniques are structural modeling, molecular mechanics, *docking* and *virtual screening*, molecular dynamics, conformational analysis and analysis of physicochemical properties. In fact, molecular modeling studies on isolated (macro)molecules or drug-receptor complexes make it possible to predict and describe the most diverse characteristics of such structures, exposing data that are close to the values determined experimentally, such as: geometric and energetic properties - lengths and angles (rotations) of atomic bonds, native three-dimensional state, intermolecular interactions, potential and kinetic energies (SANTAMARIA et al., 1998; LEACH, 2001; CHARPENTIER, 2002; ZHANG et al., 2009); thermodynamic properties - Gibbs free energy, entropy, heat of formation, heat capacity, vibration, infrared and ultraviolet spectrum (LEACH, 2001; HORN; CLARK, 2007; LANIG et al.., 2006; YEH; LEE; OLSON, 2008); electronic properties - atomic charge, total electronic density, spin density, electrostatic potential, ionization energy, molecular orbitals, frontier orbitals (LEACH, 2001; HORN; CLARK, 2007; CLOSE, 2011). The use of modeling techniques with efficient algorithms (acceptable computational cost), which have a good level of accuracy compared to experimental data, has made it possible to calculate increasingly complex and comprehensive molecular structures, especially geometry optimization calculations (CAPRASECCA et al., 2014; HABIB et al., 2015).

In medicinal chemistry, specifically in the rational planning of bioactive compounds, molecular modeling has established itself as a powerful tool in the process of identifying, selecting, manipulating, optimizing and characterizing new drugs, due to its ability to determine thermodynamic, energetic and structural properties of drug-ligand systems (BARREIRO; FRAGA, 2001; RUSSO, 2002; RAHA et al., 2007; ZHOU; HUANG; CALFLISCH, 2010). Regarding this last aspect, quantum investigations of drug-receptor binding affinity are becoming increasingly important and popular in drug research, especially those that quantify the individual contributions of each receptor residue to the total interaction energy of the system, which allow the design of new ligand derivatives with more efficient inhibitory action, resistance to mutations and with fewer adverse effects (ZHOU; HUANG; CAFLISCH, 2010; MENIKARACHCHI; GASCÓN, 2010).

In the field of nanotechnology, many studies have explored the electronic properties of DNA and oligopeptides, with the aim of using them in nanodevices (ALBUQUERQUE et al., 2014). Bezerril et al. (2009) demonstrated the dependence of charge transport through DNA on the sequence of its primary conformation using calculations of transmittance and current in Fibonacci, Rudin-Shapiro, random and Ch22 structures. In another study by the same group, the profiles of the *IxV* curves of the 5Q and 7Q variants of the a₃-peptide showed that the tunneling current in the 5Q fibrous structure is much higher than in the 7Q variant (BEZERRIL et al., 2011). In thermal terms, the substitution of alanine for glutamine does not seem to affect the specific heat and chemical potential properties of the a₃-peptides, as it does with the electronic properties (MENDES et al., 2012).

Molecular modeling is a chapter of the natural sciences that has evolved considerably in recent years, so much so that it is now integrated into much of physics, chemistry and genetics research. The expansion of this area of knowledge is due, among other factors, to conceptual advances in theoretical chemistry, especially in the field of quantum mechanics; the rapid development of computational resources; as well as the constant evolution of the functionalities of modern software for analyzing molecular structures, which allow for a complex electronic and thermodynamic characterization of the models studied (BULTINCK; TOLLENAERE; WINTER, 2003; ANDRICOPULO; MONTANARI, 2005; JÓNSDÓTTIR; JORGENSEN; BRUNAK, 2005; GUIDO; ANDRICOPULO; OLIVA, 2010; YAN et al., 2011). In fact, the existence of high-performance computational *clusters* and the increased accuracy of theoretical methods have made it possible to complement experimental data with those generated by modeling in response to biological questions (MA; NUSSINOV, 2004).

Traditional classical and quantum tools are accurate in describing the structure, physical and chemical properties of small molecules. Even today, existing *hardware* technology limits the accuracy of studies of molecules with more than 200 atoms (RAHA et al., 2007). In fact, only models between 20-200 atoms, which corresponds to the ligand and up to 20 amino acids of the receptor, can be explicitly studied in a purely quantum way (HU et al., 2009). However, nowadays, the high computational cost of dealing with a macrosystem is outweighed by separation and slicing techniques - *Fragment-based Methods.* Widely used in drug-receptor complexes, the central idea of these methods is to divide the bioreceptor into a series of small fragments in order to predict the properties of the system as a whole, based on conventional quantum (energy) calculations directed at its subsystems. And thus, for example, quantify and compare the binding energies of various ligands in the same bioreceptor, allowing the identification of the drug that interacts most strongly among all those marketed by the pharmaceutical industry (da COSTA et al., 2012; GORDON et al., 2012).

In view of the above, it can be seen that the use of molecular modeling methodologies makes it possible to generate knowledge about structural factors, geometric and energetic parameters, and electronic properties that are fundamental to the study of isolated molecules or even complex macromolecular systems. In this book, quantum molecular modeling techniques were applied (1) to study the internal interaction energies of a complex biological model and, consequently, its conformational stability; and (2) to study the electronic properties of oligopeptides and nucleic acids.

In this work, the interaction energies between the amino acid residues of T3-785, a triple-helix collagen model, were estimated. This data was then used to characterize the attractiveness of the three α chains that make up the homotrimeric peptide, as well as the stabilizing effect of each Gly-Xaa-Yaa triad that makes it up. To do this, quantum mechanics calculations were carried out within the scope of Density Functional Theory (DFT) using the Molecular Fractionation with Conjugated Hoods (MFCC) technique. As can be seen, this essentially descriptive study not only exposed but also elucidated various structural peculiarities that are essential for the integrity of the triple-helix state of collagen.

2 QUANTUM DESCRIPTION OF THE STABILITY OF HUMAN COLLAGEN

2.1 Global perspective on collagen

The term "collagen" describes a superfamily of structurally related proteins located mainly in the extracellular matrix of different types of connective tissues, in which they play an essential role in maintaining their strength and structural integrity (ASZÓDI et al., 2006). In vertebrates, collagen is the main constituent of this region, to the point of being the most abundant protein in mammals - corresponding to approximately 1/3 of the total protein mass (SHOULDERS; RAINES, 2009; Di LULLO et al., 2002). The collagen family is made up of 28 members, whose alpha-helix chains differ in the number and order of amino acid residues, the way in which they are associated with each other, interruptions in the helices and differences in the termination of the helical domains (SANDHU et al., 2012; RICARD-BLUM, 2011).

For many decades, researchers have sought to understand in molecular terms the mechanisms that contribute to the folding and stability of the collagen triple helix. Knowledge of these factors would provide valuable information on (1) the maintenance and growth of connective tissues and related pathologies, and (2) the biological activities mediated by these molecules, such as cell signaling and interaction with components of the extracellular matrix (metalloproteins, integrin). However, the conformational complexity of the collagen structure, even with highly repeated amino acid sequences, and its fibrillar nature, make biophysical and biochemical investigations of the whole peptide impossible (FIELDS; PROCKOP, 1996). For this reason, a reductionist approach, whereby smaller molecules serve as models for predicting the properties of larger, structurally related systems, has been used since the 1950s (RAMACHANDRAN; KARTHA, 1954). With this strategy, various crystallographic models of collagen-like peptides have been synthesized and their structural aspects analyzed, which has made it possible to understand many of their biochemical characteristics and their strong conformational stability (BELLA et al., 1994; BELLA; BRODSKY; BERMAN, 1995; KRAMER et al., 1998; KRAMER et al., 1999; KRAMER et al., 2000; KRAMER et al., 2001; BERISIO et al.., 2000; VITAGLIANO et al. 2001; BERISIO et al., 2002; OKUYAMA et al., 1981; NAGARAJAN; KAMITORI; OKUYAMA, 1999; OKUYAMA et al., 2004; JIRAVANICHANUN et al., 2005; JIRAVANICHANUN; NISHINO; OKUYAMA, 2006; OKUYAMA et al, 2007; EMSLEY, 2000; SCHUMACHER; MIZUNO; BACHINGER, 2006; SCHUMACHER; MIZUNO; BACHINGER, 2005; KAWAHARA ET AL., 2005; HONGO ET AL., 2001; OKUYAMA et al., 2004; HONGO et al., 2005; OKUYAMA et al., 2011; OKUYAMA et al., 2012).

In fact, for many years, excellent reviews have dealt with the use of models to elucidate the structure and function of collagen (RICARD-BLUM, 2011; BHOWMICK; FIELDS, 2013; BERISIO, 2009; FIELDS, 2010; BRODSKY; PERSIKOV, 2005). With this same purpose, specifically in section 2.1 of this chapter, aspects related to: (1) collagen diversity and biosynthesis; (2) its biotechnological applications; (3) molecular structure and stability; (4) genetic disorders; and, finally, (5) some specificities of the collagen model evaluated in this book will be addressed; the special relevance of the latter two can be highlighted in advance.

2.1.1 Diversity and functions

The collagen superfamily extends throughout the connective tissues of vertebrates and is made up of 28 types of molecules consisting of three α polypeptide chains with a main triple helix domain (RICARD- BLUM, 2011). Table 1 describes the different types of human collagen, their isoforms, genes of origin, localization, as well as their supramolecular and functional aspects.

The diversity of collagen molecules is due to the existence of 46 α-chains, some molecular isoforms and supramolecular structures for the same type of collagen, the existence of different promoters in the collagen gene, including the occurrence of alternative *splicing* (RICARD-BLUM, 2011). They are therefore subdivided into three subfamilies depending on the conformational organization visualized by electron microscopy: fibril-forming collagens (types I, II, III, V, XI, XXIV and XXVII), those associated with the surface of collagen fibrils (types IX, XII, XIV, XV, XVI, XIX-XXII and XXVI) and those that form three-dimensional networks of associations (types IV, VI, VII, VIII and X).

As can also be seen in Table 1, collagen molecules exist in practically every part of the human body, mainly in the dermis (types I, III, V, VI, VII, VIII, XII, XIII, XIV, XVI, XXVIII), basal membrane (types IV, VII, XVIII, XVIII, XIX, XXVIII), cartilage (types II, IX, X, XI, XIV, XXVII), tendons (types I, XII), ligaments (types I, XII), bones (types I, V, VI, XIV, XXIV), eyes (types II, V, VI, IX, XX, XXIII, XXIV), tissue junctions (type XXII) and connective tissues of the digestive tract (types XVIII, XXI), circulatory tract (types III, XV, XXVIII), cardiac tract (types VIII, XIII, XV, XXIII, XXV), urinary tract (types VII, VIII, XV, XVI, XXI) and reproductive tract (types XV, XXV, XXVI).

Table 1 - An overview of known human collagens

Types [chains and isoforms]	Genes	Nature Supramolecular	Location	Pathology	Functions
I [al(I)+2a2(I)]	COLIAI, C0L1A2	fibril former	diffuse - dermis, bone, tendon, ligament	osteogenesis imperfetta, ehlers-danlos syndrome, osteoporosis	tensile strength
II I3al(II)]	C0L2A1	fibril former	hyaline cartilage, vitreous humor	osteoarthritis, Chondrodysplasia	pressure resistance
III pal(III)]	C0L3A1	fibril former	skin, reticular connective tissue and smooth muscle	ehlers-danlos syndrome, arterial aneurysm	maintaining the structure of delicate tissues and
IV / 2al[IV / a2[IV] or a3(IV)a4(IV)a5(IV)or 2aS(IV)a6(IV)}	COL4A1, COL4A2, COL4A3, COL4A4, COL4A5, COL4A6	forming a three-dimensional network (tetramer + hexamer)	basal laminae	alport syndrome	support for delicate structures, filtration
V [3αl(V) or 2al(V)a2(V) or al(V)a2(V)a3(V)J	COL5A1, COL5A2, COL5A3	fibril former	diffuse - dermis, bone, cornea, placenta	ehlers-danlos syndrome	tensile strength
VI [al(VI)a2(VI)a3(VI) or al(VI)a2(VI)a4(VIf]	C0L6A1, COL6A2, COL6A3	forming a three-dimensional network	diffuse - bone, cartilage, cornea, dermis	bethlem myopathy	-
VII pal(VII)a2(VIIft	C0L7A1	Anchor fibrils	dermis, basement membrane, urinary bladder	inflammatory epidermolysis bullosa acquisita	binds connective tissue cells
VIII pal(VIII) or 3a2(VIII) or 2al(VIII)a2(VIIIJ]	C0L8A1, COL8A2	forming a three-dimensional network	diffuse - dermis, brain, heart, kidneys	fuchs endothelial dystrophy	-
IX [al(IX)a2(IX)a3(IXJ] ×[3al(X)]	COL9A1, COL9A2, COL9A3 COL10A1	Associated with type II fibrils (interrupted triple helixes) forming a three-dimensional network	Hyaline cartilage, cornea, vitreous humor hyaline cartilage	osteoarthritis, multiple epithelial dysplasia Chondoplasia	lateral association of fibrils
XI [al(XI)a2(XI)a3(XIJ]	C0L11A1, C0L11A2	fibril-forming (corn o Type II)	Hyaline cartilage, intervertebral disc	chondoplasia, osteoarthritis	pressure resistance
XII pal(XII)]	C0L12A1	Associated with type I fibrils (interrupted triple helices)	dermis, tendon, ligaments	-	lateral association of fibrils
XIII pal(XIII)]	C0L13A1	Associated with	endothelium,	-	-

XIV [3al(X/V)]	C0L14A1	membranes (interrupted triple helices)	dermis, eye, heart	-	-
XIV [3al(X/V)]	C0L14A1	Associated with diffuse - type I fibrils (interrupted triple helices)	bone, dermis, cartilage	-	-
XV [Jal(XV)]	C0L15A1	Associated with type I and III fibrils (multiple triple-helix domains and	capillaries, testicles, kidneys, heart	-	-
XVI βal(XV/)]	C0L16A1	Associated with fibrils (interrupted triple helixes)	dermis, kidneys	-	-
XVII Pal(XVII)]	C0L17A1	Associated with membranes (interrupted triple helices)	hemidesmosomes in the epithelium	epidermolysis bullosa	-
XVIII Pal(XVIII)]	C0L18A1	Multiple triple helix domains and interruptions	basement membrane, liver	knobloch syndrome	-
XIX βal(X/X)]	C0L19A1	Associated with fibrils (interrupted triple helices)	basement membrane, umbilical cord	-	■
XX [Jal(XX)]	COL20A1	Associated with fibrils (interrupted triple helices)	cornea	-	-
XXI [3al(XX/)]	C0L21A1	Associated with fibrils (interrupted triple helixes)	stomach, kidneys	-	-
XXII pal(XXII)]	COL22A1	Associated with fibrils (interrupted triple helixes)	tissue junctions	-	-
XXIII pal(XXIII)]	COL23A1	Associated with membranes (interrupted triple helices)	heart, retina	-	-
XXIV Pal(XXIV)]	COL24A1	fibril former	bone, cornea	-	-
XXV βal(XXV)]	COL25A1	Associated with membranes (interrupted triple helices)	brain, heart, testicles	formation of amyloid plaques	■
XXVI βal(XXV/)]	EMID2	Associated with fibrils (interrupted triple helixes)	testicles, ovary	-	-
XXVII βal(XXV//)]	COL27A1	fibril former	cartilage	-	-
XXVIII Pal(XXVIII)]	COL28A1		basement membrane, vascular tissue, sciatic nerve	neurodegenerative disease	-

Among the collagens capable of effectively organizing themselves into fibers, types I and III are formed in the skin in a much higher proportion than the others, which is kept practically constant in healthy tissues. However, this abundance is altered when scar tissue is formed as a result of old age, damage by chemical or biological agents, as in the case of infectious diseases (CHENG et al., 2011b).

Collagens of this nature are formed from a complex process involving a series of post-translational modifications at intracellular and then extracellular level. The following two paragraphs provide a general explanation and Figure 1 shows a summary of the post-translational processes.

Figure 1 - Intracellular and extracellular events in the formation of a   fibril.

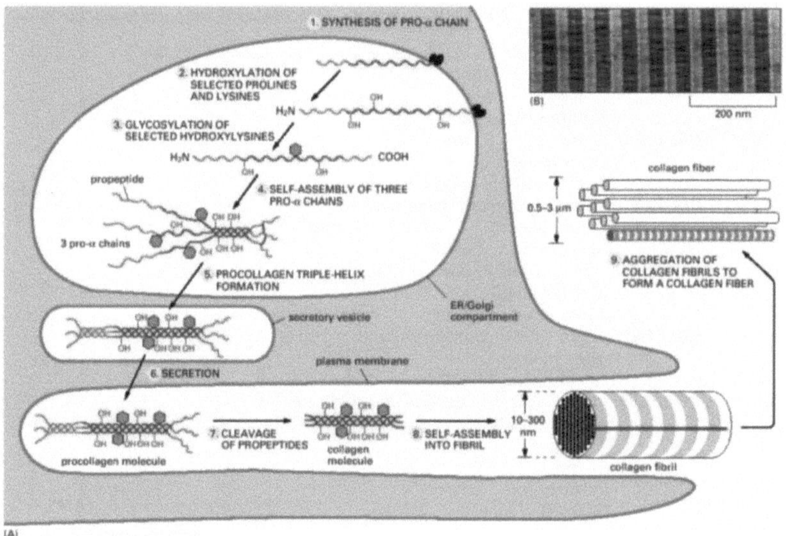

As an example, let's take the gene for the α-chain of homotrimeric type III collagen, symbol COL3A1 (NCBI Reference Sequence: NG_007404.1), located at position 2q32.2 on chromosome 2, with a length of 38,374 bp (region: 5001..43374) and 51 exons. Initially, the gene transcription process generates an mRNA with 5490 bp (NCBI Reference Sequence: NM_000090.3) which will be translated in the form of an α prepro-protein (or pro-α chain) with 1466aa (NCBI Reference Sequence: NP_000081.1) in ribosomes attached to the RER membrane. Translocation to the RER lumen occurs after cleavage of the signal peptide (1-23) by the signal peptidase. The product generated is called type III procollagen, which consists of: "N-terminal globular propeptide" + "non-helical N- telopeptide" + "central triple helix domain" + "C- telopeptide" + "C-terminal globular propeptide". The procollagens

10

(still individualized) undergo numerous post-translational modifications - hydroxylation of proline and lysine residues, sulphation of tyrosine residues, as well as glycosylation of lysine and hydroxylysine residues (MYLLYHARJU; KIVIRIKKO, 2004). At the same time, the chaperone proteins HSP47 and SPARC interact with three chains of type III procollagen, catalyzing the triple helix folding process in the C-terminal -> N-terminal direction. At this point, the trimeric structure is secreted by the cell and the propeptides are partially cleaved by N-proteinases, ADAMTS-2, and C-proteinases, BMP-1.

After the cleavage processes, the molecule formed, now called tropocollagen, is predominantly in a triple helix state and has a length (diameter) of approximately 300nm (1.5nm). During a characteristic process of self-organization (GAUTIERI et al., 2011), several tropocollagen molecules cross-link with each other through the N-/C- telopeptides, forming fibrils with diameters of 100-500nm. The collagen fibrils are organized into elongated fibres on the micrometric scale (diameter of ≈10µm) and other supramolecular patterns, which will finally constitute the so-called collagenous tissues (Figure 2). It can therefore be seen that the tropocollagen is the fundamental unit of these macroscopic fibers (networks).

Figure 2 - Hierarchical structure of fibrous collagen.

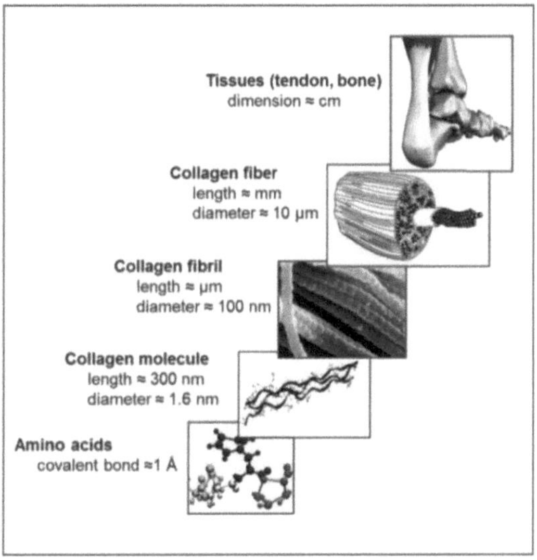

Source: *VESENTINI; RADAELLI; GAUTIREI, 2010*

As we have seen, collagenous tissues are nothing more than hierarchical structures made up of self-organized collagen molecules. Depending on the alignment of these monomers and the chemical microenvironment, these tissues have different mechanical and biological functions (CEN et al., 2008; VESENTINI; RADAELLI; GAUTIREI, 2010). For example, parallel-organized collagen fibres and a greater amount of minerals in the middle give bones rigidity; while thick, parallel-organized collagen fibres embedded in a smaller amount of minerals give tendons greater resistance to tension - not elasticity.

Despite advances in the biosynthetic dynamics of collagen, it is important to note that this process is far from being completely unraveled, due to the complexity of collagen metabolism: the process of transcription, mRNA *splicing*, intracellular transport and secretion, trimerization of propeptides, N-glycosylation and other post-translational modifications (MIZUNO et al., 2013).

In functional terms, collagen acts directly in maintaining the structural and functional integrity of the extracellular matrix of soft and hard connective tissues, due to its ability to interact with a multitude of chemical molecules diffused there (ASZÓDI et al., 2006). Because it is such an important biomolecule, many of its organic functions are currently well understood (RICARD-BLUM, 2011; SANJEEV, 2014): (1) its ability to interact with polysaccharides of the extracellular matrix and the cytoplasmic membrane of fibroblasts due to its hydrophilicity; (2) its role in the dynamic process of re-organization and reactivity of connective tissue by interacting with fibronectin filaments; (3) its participation in the primary hemostasis process by activating platelets during the adhesion and aggregation stage, and secondary hemostasis, as an activator of coagulation factors (WANG et al., 2013); and (4) its function in the organic transport of ions, electrolytes, metabolites and drugs.

The chemoattractant and hemostatic activities of collagen mentioned in the previous paragraph are related to its ability to interact with a huge range of molecules solubilized in the extracellular matrix through its cell adhesion domains (Arg-Gly-Asp *motif,* for example) (LEE et al., 2001; YAMADA et al., 2014).

1.1.2 Biotechnological applications

The repetitive nature of the collagen structure, formed by triads of amino acids, facilitates the development of study models, as well as the synthesis, isolation and purification of large quantities of natural or synthetic collagen. All this biological material can be widely applied in the fields of Regenerative Medicine and Bioengineering due to its very unique mechanical, physico-chemical and biological properties (ABOU NEEL et al., 2013; WANG; MACNEIL, 1998).

Collagen has several properties, including: its low antigenicity and consequent high biocompatibility (PATI et al., 2012; LEE; SINGLA; LEE, 2001; PARENTEAU-BAREIL; GAUVIN; BERTHOD, 2010); its low reactivity and biodegrability (LEE; SINGLA; LEE, 2001; PARENTEAU-BAREIL; GAUVIN; BERTHOD, 2010); semi-permeability (WANG; MACNEIL, 1998); enzymatic resistance to trypsin, papain and hyaluronidase (GRANT, 2007); high thermal stability and mechanical strength, necessary to maintain its fibrillar nature when subjected to large deformations and/or temperatures - the latter being not so intense, allowing a certain degree of malleability of the molecules to be maintained (ASZÓDI et al., 2006; BUEHLER, 2008; WANG; MACNEIL, 1998).

Through the use of model peptides that are sequentially and structurally similar to collagen, new components are developed that have special properties - greater thermostability, inhibition of metalloproteinases and the ability to mimic/modulate collagen interactions with molecules of medical importance (LAUER-FIELDS et al., 2007; OTTL et al., 2000).

Super-stable collagen molecules can be synthesized through the addition of branching and functional groups (SHOULDERS; HODGES; RAINES, 2006) and thus used safely in the design of support, adhesion and cell growth meshes for low-invasive tissue medicine procedures (YAMADA et al., 2014; O'BRIEN, 2011; GLOWACKI; MIZUNO, 2008). In the field of plastic surgery, for example, collagen is used in burn victims, to fill bone or skin defects (nasolabial folds) and also for peridental treatment (GROVER et al., 2010).

In hydrogel form, it can be used as a carrier molecule in slow-release drug delivery systems, such as growth factors and cytokines (WALTERS; STEGEMANN, 2014).

Another important industrial product is gelatine. This is a compound derived from the partial degradation of collagen and is widely used in the food, cosmetics, pharmaceutical and photographic industries. The effects of collagen-based food supplements on human health are still highly debatable as the scientific literature is still quite contradictory and inconclusive, even unreliable, but there are reports of: (1) an increase in the performance of high-performance athletes (CLARK et al., 2008); (2) improved sensation in chronic joint pain (BRUYÈRE et al., 2012); (3) acceleration of the bone regeneration process (AL-MOUSILLY; ALAJELI; ABDULRAHMAN, 2014); and (4) beneficial effects in the treatment of patients with rheumatoid arthritis (BAGCHI et al., 2002).

1.1.3 Molecular structure and stability

In connective tissues, tropocollagen molecules continuously complex into microfibers and then into macroscopic, elongated transverse and longitudinal fibers, which intertwine into a compact biological web that reinforces tendons, ligaments, bones and the basal lamina (HULMES, 1992).

In structural terms, simple collagen molecules are formed by three polypeptide chains in the poly-Pro II (PPII) conformation intertwined together around a common axis and in a helical conformation (triple helix) facing to the right. PPII is formed when the residues adopt dihedral angles of approximately -55° (φ), 150° (ψ), - 120° (Ω) and trans peptide bonds (RICH; CRICK, 1961; MORADI, 2009). Depending on the primary sequence, they are classified as homotrimeric - three identical α-chains (α1+α1+α1) - or heterotrimeric - two (α1+α1+α2) or three (α1+α2+α3) distinct α-chains (GORDON; HAHN, 2010). Furthermore, depending on their quaternary structure, they can be grouped into subfamilies: fibrillar; associated with fibrils; network; transmembrane; and membrane associated with interrupted triple helices (COLE, 1994) - see Table 1.

The different types of collagen, even with their structural and functional peculiarities, are formed by repeats of the *Gly-Xaa-Yaa* sequence (triad). The *Gly* residues are essential for the formation of the triple helix, as they alone are small enough to occupy the core (interior) of the helix without a strong steric hindrance (BHOWMICK; FIELDS, 2013) - Figure 3. In addition, there are alternating hydrogen bonds between the amine group of glycines (*Gly*) and the ketone group of proline (*Pro*) which also restrict this conformation.

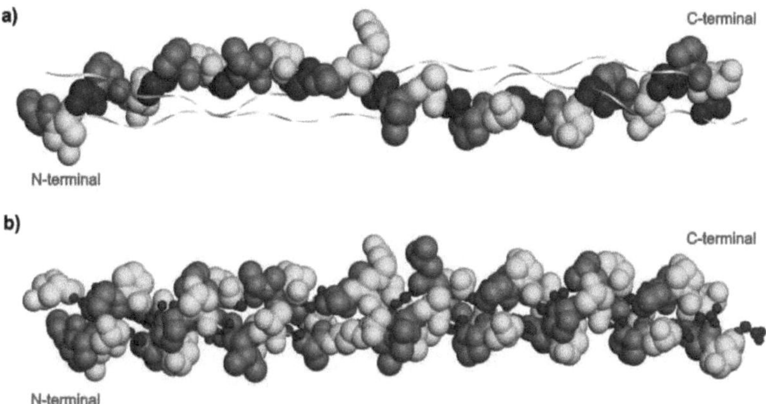

Source: *Prepared by the author*

In theory, there are more than 400 possible triads, but sequence analysis of the different types of collagen indicates that only a small number of possibilities actually occur. The *X and Y* positions are often occupied by the amino acids proline (*Pro*, 28.1%) and hydroxyproline (*Hyp*, 38.1%), respectively. Together they account for approximately 22% of all collagen ribbon residues (RAMSHAW; SHAH; BRODSKY, 1998). Other amino acids found are alanine (*Ala*), lysine (*Lys*), argenine (*Arg*), leucine (*Leu*), valine (*Val*), serine (*Ser*), and threonine (*Thr*), more rarely glycine (*Gly*), methionine (*Met*) or histidine (*His*), and never cystine (*Cys*), tryptophan (*Trp*) and tyrosine (*Tyr*) (GRANT, 2007; BHOWMICK; FIELDS, 2013).

The effects of different types of amino acids on the stability of homotrimeric systems of the *Gly-Xaa-Yaa* type were previously examined based on the *Pro/Hyp* residues in *Xaa/Yaa*. *The* most stable residues in the *Yaa (Xaa)* position were *Hyp > Arg > Met > Ile > Gln > Ala (Pro >* charged residues *> Ala > Gln*) (PERSIKOV; RAMSHAW; BRODSKY, 2000; PERSIKOV et al. 2000). These data provided the tendency/predilection of a given residue in the *Xaa* or *Yaa* position of the homotrimer, measured in terms of the destabilizing influence of each residue compared to Gly-Pro-Hyp.

From this perspective, the *Gly-Pro-Hyp* triad, as well as being the most abundant in collagen - 10.5% - can also be considered the most important. Ramshaw, Shah and Brodsky (1998) showed that helices synthesized exclusively with these triads proved to be more stable than others made up of different types of *Gly- Xaa-Yaa* sequences.

According to Dai, Wang and Etzkorn (2008), proline residues and their derivative *Hyp play* an important role in the triple helix folding of collagen. They are able to pre-organize the strands of the molecule into a PPII conformation, which reduces the entropic cost of folding. In terms of geometric isomerism, the *Pro* residues in peptide chains generally appear in *cis* and *trans* conformations (due to the formation of tertiary amides), but the residues that make up a triple helix preferentially have *trans* isomerism. Thus, before a (*Pro-Hyp-Glu*)*n* strand can fold into a triple helix, *cis→trans* isomerization of all proline residues must occur. Preventing this process with an inhibitor results in the destabilization of the triple helix, even when the model under study maintains all the interchain hydrogen bonds.

After this isomerization process, *Pro* residues in the *Yaa* position of the protocologen triads are modified by the enzyme prolyl-4-hydroxylase to form *Hyp*, commonly used as a characteristic marker of collagen (GRANT, 2007). It is believed that the hydroxylation reactions probably take place before the molecule reaches the triple-helix state (PROCKOP; KIVIRIKKO, 1984). The absence of this post-translational process directly affects the stability of collagen (PERRET et al., 2001).

In fact, the product (2S,4R)-4-hydroxyproline (*4RHyp*), when located in the *Yaa* position, plays a crucial role in the folding and conformational and thermal stabilities of the tropocollagen triple helix (MIZUNO; HAYASHI; BĂCHINGER, 2003; BERG; PROCKOP, 1973; FIELDS; PROCKOP, 1996). For example, the denaturation temperature increases from 30°C to 60°C when comparing (*Pro-Pro-Gly*)10 with (*Pro-Hyp- Gly*)10 (ROSENBLOOM; HARSH; JIMENEZ, 1973).

When *4RHyp* residues reverse their position from *Yaa* to *Xaa*, the integrity of the triple helix is compromised. Even its non-formation has been reported for the (*Gly-Hyp-Pro*)10 model (MIZUNO; HAYASHI; BĂCHINGER, 2003). Also from this perspective, Improta, Berisio and Vitagliano (2008) quantitatively showed that dipole-dipole interactions from the hydroxyl groups contribute to *4RHyp*'s preference for the *Yaa* position. In this case, the addition of a hydroxyl group in *Pro in* the *Gly- Xaa-Yaa* triad can have a stabilizing or destabilizing effect if located in *Yaa* or *Xaa,* respectively.

However, this destabilization scenario when *4RHyp* is in *Xaa* was not observed in the *Gly-Pro-Hyp* (DOI et al., 2005; BERISIO et al., 2004) and *Gly-Pro-Thr* (BANN; BACHINGER, 2000) triads. On the contrary, slight but significant pro-stabilization effects have been observed in triple-helix models. In the first case, direct or water-mediated hydrogen interactions established between adjacent *Hyp* justify this finding. An explanation for the second molecule will be presented later.

Precisely because it is a key element in collagen, some hypotheses have emerged to explain the stabilizing effects of *4RHyp*. Initially, it was suggested that water-mediated hydrogen bonds between the hydroxyl group of *4RHyp* in *Yaa* with an amide located in distinct chains promotes the stabilization of the triple helix (BELLA; BRODSKY; BERMAN, 1995; SUZUKI; FRASER; MACRAE, 1980; RAMACHANDRAN; BANSAL; BHATNAGAR, 1973; BELLA, et al, 1994). However, some studies focusing on the effect of small covalent modifications to the *Hyp* structure have indicated alternative mechanisms. Experimentally, the substitution of *4RHyp* by (2S,4R)-4-fluoroproline (*4RFlp*), but not the *4S* isomer, greatly improves the stability of triple-helix conformations (BETSCHER et al., 2001; NISHI et al., 2005) even though fluorine is unable to form strong hydrogen bonds (O'HAGAN, 2008). It has been proposed that the "O" ("F") group of *4RHyp* (*4RFlp*) favors the stability of collagen by inducing (inductive effect) the *trans* conformation of the peptide bond that precedes it.

Vitagliano et al. (2001) observed that imino acids (pyrrolidine ring) in the triple helix should preferably adopt the *gauche* conformation[+] = *"DOWN"* (dihedral angle $\chi1$ *positive, $\chi2$ negative, $\chi3$ positive and $\chi4$ negative*) and *gauche*[-] = "UP" (dihedral angle $\chi1$ negative, $\chi2$ positive, $\chi3$ negative and $\chi4$ positive) in the *Xaa* and *Yaa* positions, respectively. In the case of the pyrrolidine rings of Pro and Hyp, these conformations are also called Cγ-endo or Cγ-exo ("DOWN" or "UP") respectively.

This kind of favoring of one or the other position in the *Gly-Xaa-Yaa* triad *is* called "positional preference". As *4RHyp* commonly adopts the *gauche-* state, it consequently does not have access to the *Xaa* position, but to *Yaa*. This behavior occurs in very rigid systems, which is not the case for peptides with the *Gly-Hyp-Thr* sequence repetitively.

Bretscher et al. (2001) stated that, in fact, high electronegativity substituents, such as the "O" ("F") in position 4 on the pyrrolidine ring of the *4RHyp* residue (*4RFlp*) in *Yaa*, stabilize the collagen triple helix by means of stereoelectronic effects - *gauche effect* together with $n{\rightarrow}\pi^*$ interaction - and not a simple inductive effect. It is important to remember that the *gauche effect* is related to *4RHyp*'s preference for the *gauche* conformation⁻ .

4RHyp's conformational preference for the *gauche state*⁻ stabilizes the *Cγ-exo* (*up state*) of the respective pyrrolidine ring. When this conformational state is reached, a $n{\rightarrow}\pi^*$ interaction stabilizes the *trans* isomer of the peptide bonds. Both effects pre-orient the dihedral angles of each oligopeptide chain in a favorable conformation for the formation and preservation of the triple helix - Figure 4.

Figure 4 - Stereoelectronic effects that stabilize the triple helix of collagen: (a) *gauche* effect and $n{\rightarrow}\pi^*$ interaction capable of re-organizing the torsion angles and, consequently, promoting stability of the triple helix. (b) the *gauche effect* promoted by the R group[1] = EWG=OH/F in the 4 *R* position (proportional to the electronegativity of this element), stabilizes the Cγ-exo conformation of the modified pyrrolidine ring and no longer the Cγ-endo (the natural preference of prolines). (c) the $n{\rightarrow}\pi^*$ interaction, which favors the *trans* isomer of the peptide bond, especially when the proline derivatives are in the Cγ-exo state; (d) the overlap of the natural n and π^* orbitals of a Pro in Cγ-exo, which accentuates the interactions between adjacent polar groups.

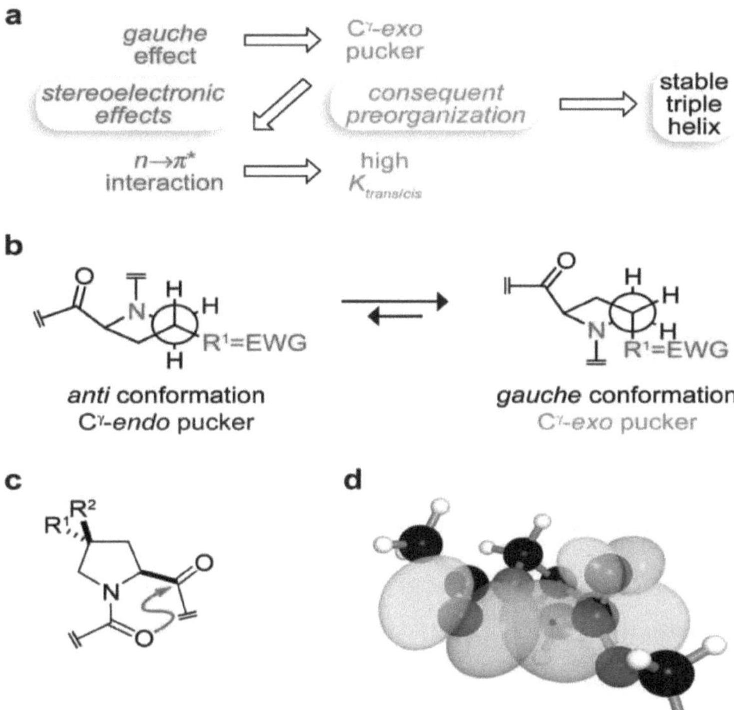

Source: *SHOULDERS; RAINES, 2009.*

To highlight the importance of stereoelectronic effects, Kotch, Guzei and Raines (2008) studied the effect of simple covalent modifications (in this case, O-methylations) on the structure of *Hyp* that were able to preserve the stereoelectronic effect of the hydroxyl group. They observed that the (2S,4R)-4-methoxyproline residue occupying the *Yaa* position (i.e. replacing *Hyp*) in (*Pro-Hyp-Gly*)₁₀ confers greater stability to the triple helix even though it reduces the local solvation layer. Due to similar stereoelectronic effects, the *Cγ-exo* conformation of the 4-substituted ring of the *Hyp* residue is favored in (2S,4R)-4-methoxyproline (KOTCH, 2008) and 4R-fluoroproline (BETSCHER et al., 2001; NISHI et al., 2005). Thus, both studies show that *Hyp*'s hydroxyl group acts primarily through stereoelectronic effects. However, Shoulders and Raines (2011) in their studies of the *(Hyp-Hyp-Gly)*n structure concluded that interchain dipole-dipole interactions between *Hyp* residues are the main determinants of the hyperstability characteristic of these systems.

From the above, it can be seen that there has been much discussion about the individual factors capable of explaining the importance of the different types of residues that make up the structure of collagen. The role of hydrogen bonds and other inter/intra-chain dipole-dipole interactions versus inductive effects (of a stereoelectronic nature) in stabilizing the triple helix of collagen is one of the targets of this debate.

In this context, the study by Hodgers and Raines (2005) demonstrated the possibility of obtaining heterotrimeric helices from the association of peptides that are individually incapable of forming stable helices. This work strongly highlighted the role and importance of interchain interactions for the triple helix state.

Buehler (2008) demonstrated, through collagen stress-strain curves, that the formation of covalent cross-links between tropocollagen units is closely related to the ability of collagen to undergo deformation without rupture, either due to elastic or even dissipative behavior. An increase in the density of cross-links leads to hardening and, consequently, an increase in the fragility of collagenous tissues (BAILEY; SETHNA, 2003; BUEHLER, 2006).

1.1.4 Genetic disorders

The stability of collagen is highly sequential, i.e. it depends on the integrity of the type and order of the amino acid residues encoded. Therefore, structural mutations (substitutions, deletions, duplications and insertions) in the genes that code for the different types of human collagen can cause a series of genetic disorders of connective tissue, such as Osteogenesis Imperfecta (type I), Chondrodysplasia, Ehlers-Danlos syndrome, inflammatory epidermolysis bullosa acquisita (type VII) and hypertrophic achondroplasia. Even simple mutations can lead to biochemical and clinical disorders with extremely deleterious phenotypic consequences (BYERS; COLE, 2002).

Marlowe, Singh and Yingling (2012) conducted a series of molecular dynamics simulations to assess the effect of point mutations on the structure of collagen and its mechanical properties. It was commonly observed that mutations destroy the structure and reduce the strength of collagen fibrils, which can affect their packing.

As is well known, the presence of the *Gly* residue every three amino acids is a prerequisite for the formation of collagen fibrils. Therefore, substitution of a single residue for another usually induces some pathological condition. This is the case with osteogenesis imperfecta, chondrodysplasia and *ehlers-danlos* syndrome, caused by changes in type I, II and III collagen, respectively. Thus, it is believed that specific substitutions of *Gly* residues in natural collagen induce distortion and destabilization of the triple helix with a deleterious effect on human health.

In 1994, Bella et al. found that the *Gly→Ala* substitution results in a subtle change in conformation, with a local unwinding of the triple helix. At the mutation site, interchain hydrogen bonds were broken and replaced by others mediated by interstitial water. A recent study of structural refolding after subjecting the system to high temperatures indicates that the *Gly→Ala* mutation can be easily compensated for and the renucleation process quickly normalized after the mutated site (MIZUNO, 2013), which is not the case with the *Gly→Val* modification. In contrast, modifications of the *Gly→Arg* or *Gly→Ser* type in recombinant collagens led to a slight decrease in stability (CHENG, 2011a).

Ehlers-Danlos Syndrome (EDS) type IV, commonly called vascular EDS because it induces spontaneous rupture of the great vessels, is an autosomal dominant disease caused by mutations in the COl3A1 gene on chromosome 2q31, which codes for the alpha-1 chain of human type III collagen. As a rule, these point mutations result in thermal destabilization and slowing down of the folding and secretion process of the triple helix, so as to generate certain abnormalities in the post-translational processing of collagen, such as superglycosylation of hydroxylysines and stimulation of the degradation system (SUPERTI-FURGA et al., 1989; SUPERTI-FURGA; STEINMANN, 1988; SUPERTI-FURGA et al., 1988; NUYTINCK, et al., 1992; MIZUNO et al., 2013; BONADIO; BYERS, 1985).

Since the 1980s, molecular biology techniques such as PCR and electrophoresis have been used to identify and investigate mutations in the COL3A1 gene. In fact, in 1988, Professor Superti-Furga's group identified the first mutation related to the syndrome, a 3.3 kb deletion in the coding domain of one of the two COL3A1 alleles (SUPERTI-FURGA, 1988b). Since then, 621 amino acid substitutions have been confirmed and are currently signed in *The Human Type I and Type III Collagen*

Mutation Database (DALGLEISH, 1998). Approximately half of these are *Gly* residues, which alone account for 28.6% of the amino acids in the type III collagen proprotein (NCBI Reference Sequence: NP_000081.1). According to Mizuno et al. (2013), the substitution of a single nucleotide for *Gly* can cause nine possible mutations ($1/4$ for *Arg*, $1/6$ for *Ala/Val*, $1/12$ for *Asp/Glu/Ser/Cys*, $1/24$ for *Trp*, and $1/24$ for a termination codon).

Examples include the substitutes *Gly847Glu* (FOX et al., 1988; RICHARDS et al., 1992), *Gly571Ser* (GILCHRIST et al., 1999), *Gly883Val* (PALMERI et al., 2003), *Gly1018Asp* (KONTUSAARI et al., 1992), *Gly1006Glu* (JOHNSON et al., 1992), *Gly1021Glu* (NARCISI et al., 1993), *Gly136Arg* (TROMP et al., 1993), *Gly637Ser* (NARCISI et al., 1994), *Gly499Asp* (MCGRORY; COSTA; COLE, 1996), *Gly415Ser* (ANDERSON et al., 1997), *Gly934Glu* (MCGRORY et al., 1996), *Gly571Ser* (GILCHRIST et al., 1999), *Gly16Ser* (PEPIN et al., 2000), *Gly82Asp* (PEPIN et al., 2000), *Gly373Arg* (PEPIN et al., 2000), *Gly385Glu* (PEPIN et al., 2000), *Gly297Ser* (KROES; PALS; van ESSEN, 2003), *Gly883Val* (PALMERI et al., 2003).

Tromp et al. (1989) observed that the *Gly790Ser* substitution induces molecular destabilization characterized by the exposure of an adjacent protease-sensitive site after a local unfolding of the triple helix.

Thus, the *Arg* residue at position 790, which would naturally undergo trypsin action at 30-35°C, starts to react even at low temperatures (0, 10 and 20°C) (STOLLE et al., 1985; TROMP et al., 1989).

In 1995, TROMP et al. identified an important mutation, *Gly793Val*. A recent study shows that *Gly→Val* mutations, besides being more common, are much more severe when compared to *Gly→Ala*. By inducing a sharp drop in the folding rate of collagen, this mutation promotes super-glycosylation of hydroxylysine residues and may even stimulate the degradation system (MIZUNO et al., 2013).

2.2 Importance of the work

Motivation

According to the global perspective discussed in the previous section, it can be inferred that the collagen superfamily is made up of structurally related peptides that are widespread in the extracellular matrix of vertebrates and even invertebrates, where they contribute to the physical and biological properties of the tissues of these immulticellular organisms (EXPOSITO; LETHIAS, 2013). Characteristics related to its reactivity, thermal stability, mechanical and enzymatic resistance, biocompatibility, low antigenicity, biodegrability and its ability to be involved in specific interactions with other biomolecules, make collagen fibers important for so many biological organisms and for the biotechnology industry (PATI et al., 2012; LEE; SINGLA; LEE, 2001; PARENTEAU-BAREIL; GAUVIN; BERTHOD, 2010; ABOU NEEL et al., 2013). Therefore, understanding how such properties derive from their fundamental structural units requires a comprehensive knowledge of the mechanisms underlying their structure and stability (SHOULDERS; RAINES, 2009).

Although many theoretical and experimental studies have aimed to describe the stability of this molecule, few quantify the binding affinity between its amino acid residues. Only in 2008, for example, were the interactions between the residues that make up a particular triad of a collagen model energetically quantified and evaluated to justify the stability of the triple helix (IMPROTA; BERISIO; VITAGLIANO, 2008).

Hypothesis

The high stability of the collagen triple helix can be characterized and quantified in terms of energy. This includes identifying the residues and triads that act most effectively in this stabilization.

General objective

Analyze the conformational stability of collagen, using an effective quantum method capable of describing the interaction energies between the amino acid residues of the *A*, *B* and *C* chains of a model system with high structural similarity (T3-785, PDB:1bkv).

Specific objectives

S Quantify the residue-residue (i.e. pairwise) interaction energies of the amino acids that make up T3-785, in the order - residues of chain *A* interacting with residues of chain *B* (subsystem *AB*), residues of chain *B* interacting with residues of chain *C* (subsystem *BC*), residues of *C* with *A* (subsystem *CA*) and joint interactions added together (complex *ABC*);

S Explain the intensity of the residue-residue interaction energies based on the chemical nature of the intermolecular forces between the pairs;

S To investigate the energetic attractiveness of the N-terminal, central and C-terminal zones of the *A*, *B* and C chains in the *AB, BC* and *CA* subsystems, respectively;

S To investigate the energetic attractiveness of the *Pro-Hyp-Gly, Ile-Thr- Gly, Ala-Arg-Gly* and *Leu-Ala-Gly* triads of the *A*, *B* and *C* chains in the *AB, BC* and *CA* subsystems, respectively;

S Analyze the relevance (importance) of the individual amino acids *Pro, Hyp, Gly, Ile, Thr, Ala, Arg* and *Leu* for the stabilization of the triple helix model;

S Evaluate the energetic impact in terms of stability of point substitutes on the wild structure of collagen.

2.3 Description of the collagen model analyzed

In this book, the consistency of the triple helix of human collagen was evaluated on the basis of an extensive study of the energy forces existing in the synthetic peptide T3-785, whose three-dimensional structure was determined using X-ray diffraction at a resolution of 2.0 Å and stored in the *protein data bank* under the code 1BKV (KRAMER et al., 1999). This peptide presents the first real crystallographic view of an imino acid-free region of human collagen. Specifically, it contains three identical chains of a 12-residue sequence (785-796) of amino acids common to Type III collagen, of which the *Ile-Thr-Gly, Ala-Arg-Gly* and *Leu-Ala-Gly* triads stand out. These occupy the central region or zone of T3-785 and respect the distribution observed in other types of collagen: *Arg* and *Thr* preferentially occupy the *Yaa* position, and *Leu,* the *Xaa* position (RAMSHAW; SHAH; BRODSKY, 1998). In addition to this central region, the T3-785 molecule has two terminal zones (C- and N- terminal) formed by repeats of the *Gly-Pro-Hyp* sequence, which is known to provide stability to the helices (high melting point) when in neutral pH and aqueous media (RAMSHAW; SHAH; BRODSKY, 1998) - Figure 5. For an individual view of all the residues that make up T3-785 and the respective water molecules that surround them, see Figures 6, 7 and 8.

Figura 5 - (a) schematic, (b) cross-sectional and (c) longitudinal representation of the three zones that structurally make up the T3-785 peptide.

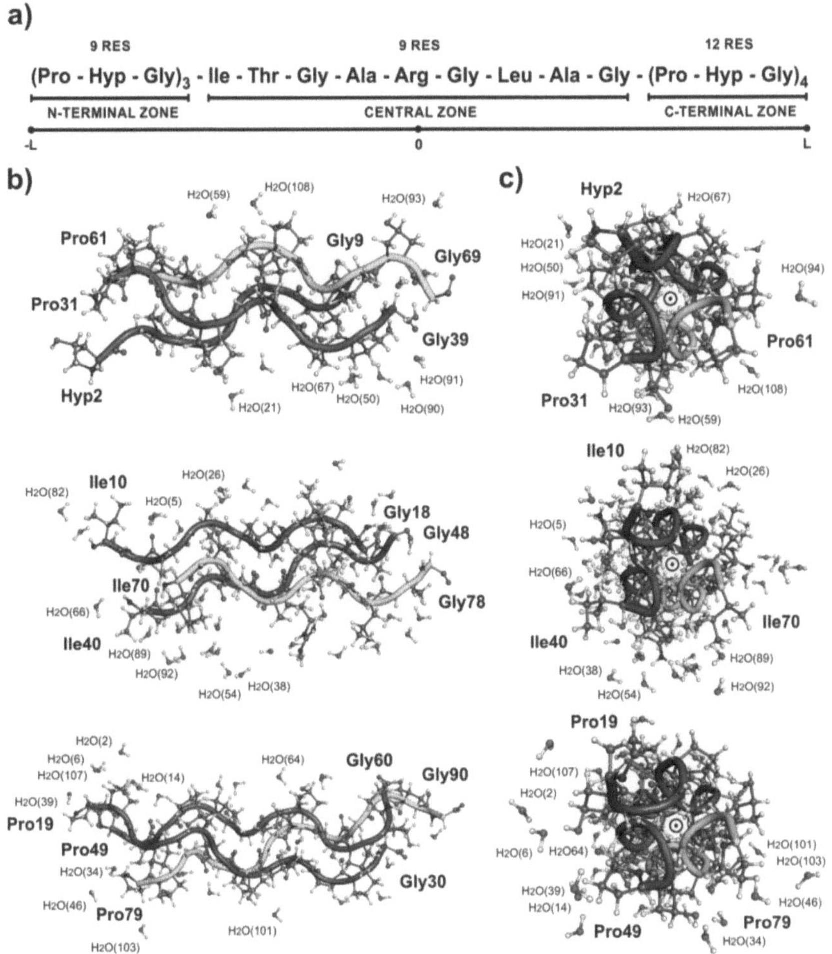

Source: *Prepared by the author*

Figura 6 - *Amino* acid residues from the *A-chain of* T3-785, together with their surrounding water molecules.

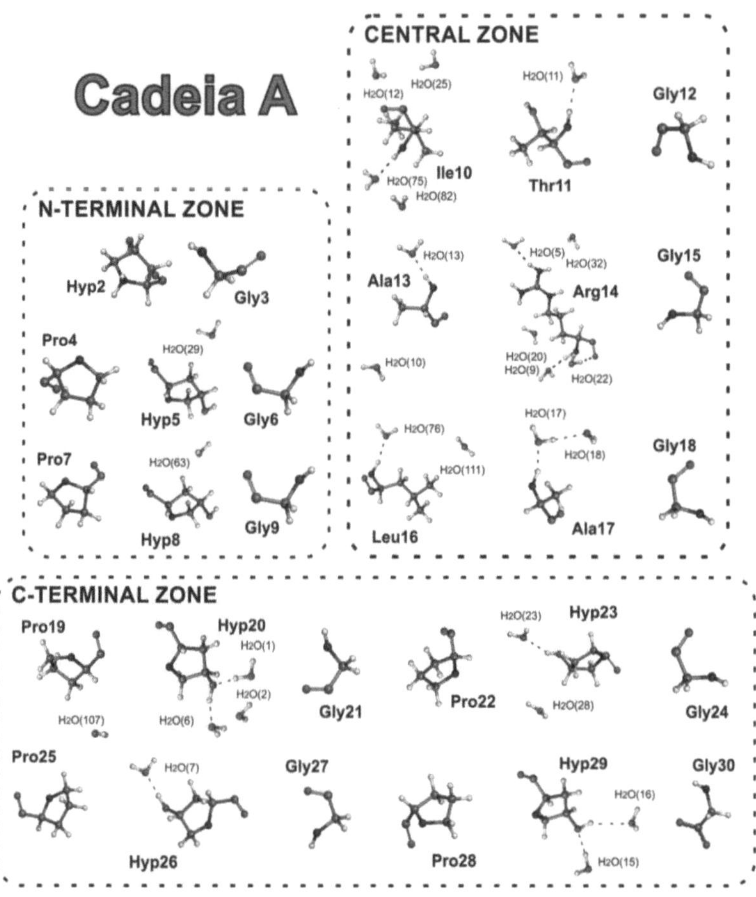

Source: *Prepared by the author*

Figura 7 - Residues of amino acids from the *B* chain of T3-785, together with their surrounding water molecules.

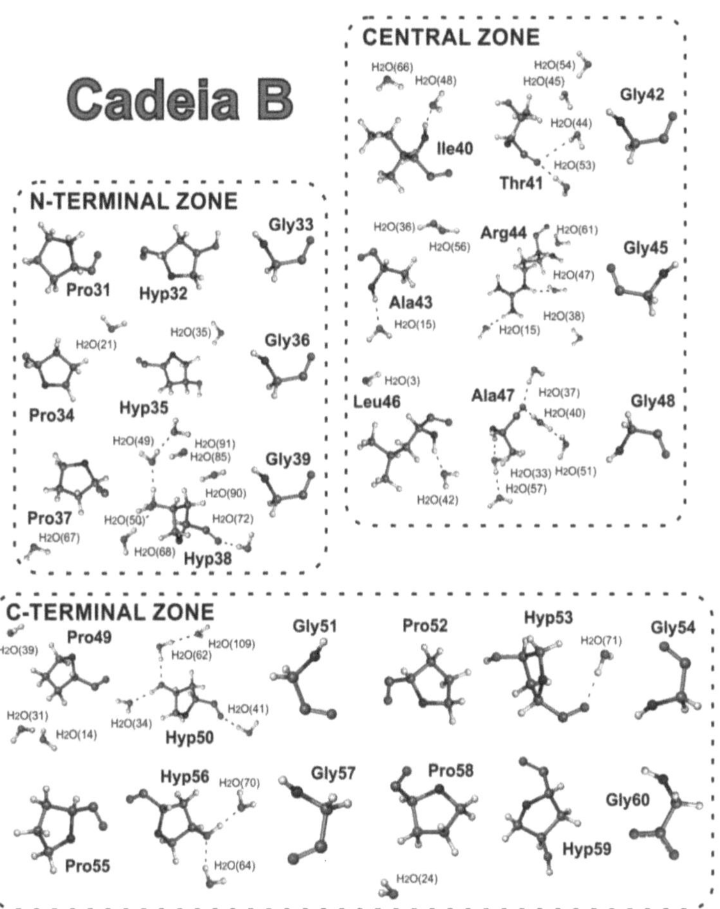

Source: *Prepared by the author*

Figura 8 - Residues of amino acids from the *C-chain* of T3-785, together with their surrounding water molecules.

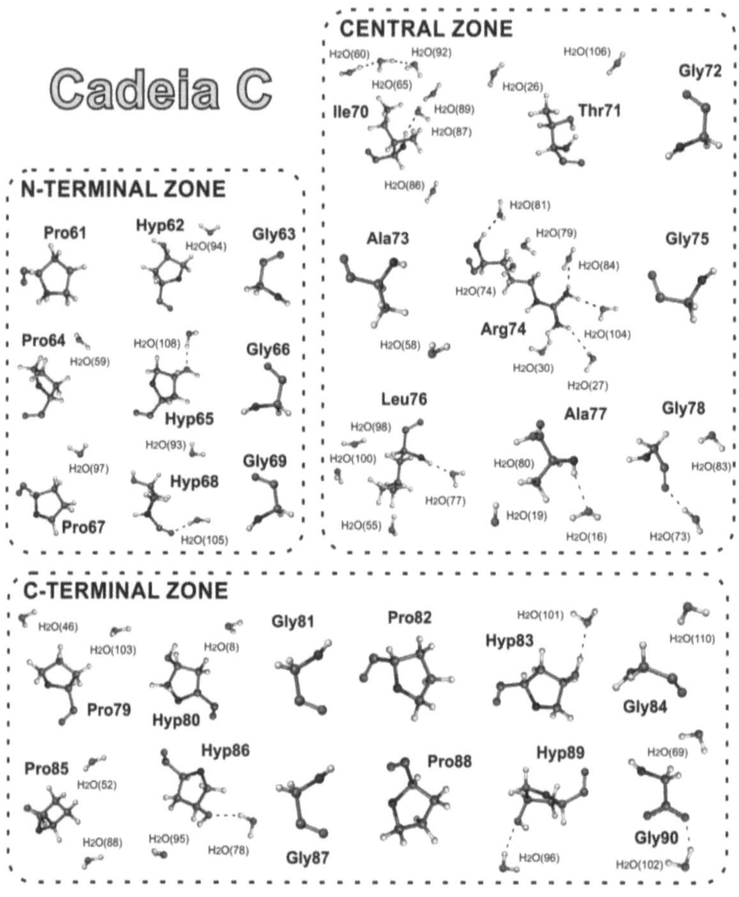

Source: *Prepared by the author*

Diffraction patterns of the collagen fiber are typically interpreted assuming the triple-helix conformation of the individual molecule. Two symmetries are observed: the 7/2 (*7-fold*) symmetry, in which seven residues are located every two turns; and the 10/3 (or *10-fold*) symmetry, characterized by the presence of 10 residues every three turns. Analysis of crystallographic structures suggests that natural collagen gradually and continuously varies its helical twist according to the local distribution of imino acids (BELLA, 2010). In general, regions rich in imino acids prefer to adopt the 7/2 helix pattern, due to the restriction of the geometry of the proline ring. This symmetry is broken by the relaxation in the structure promoted by the variability of Φ and Ψ angles possible for different *Pro* and *Hyp* residues (OKUYAMA et al., 2012). In the case of the T3-785 peptide, 7/2 symmetry predominates in the terminal regions, while the predominant symmetry in the region lacking Hyp/Pro (central region) is 10/3 (KRAMER et al., 1999).

This double symmetry alters the distance between the residues of adjacent chains, as can be seen, for example, when *Arg14* (the 14th element in chain *A*) interacts with residues in chain *B* that occupy the 11[a] to 17[a] position. However, the same is not observed when *Arg74* (14th element in the *C chain*) interacts with residues in the *A chain* that occupy the 14[a] to 20[a] position.

In their native conformation, collagen molecules require hydration to maintain their triple-helix structure. Crystallographic data of peptide models similar to collagen indicate a repetitive and ordered arrangement of water molecules in the solvent-exposed regions (*Xaa* and *Yaa*) (HONGO et al., 2001; OKUYAMA et al., 2004; HONGO et al., 2005; OKUYAMA et al., 2011; OKUYAMA et al., 2012). A total of 111 water molecules are found in the T3- 785 peptide crystal, many of which make up the first and second solvation layers, capable of forming hydrogen bonds with the peptide or other water molecules, respectively.

It is a fact that hydrogen bonds are extremely important intermolecular interactions for the structural integrity of biological systems. In the T3-785 peptide, all the *Gly-Xaa-Yaa* triads (where *Xaa=Pro* and *Yaa=Hyp*) in the terminal regions have hydrogen bonds in the *Rich and Crick II* pattern (RICH; CR*I*CK, 1961), i.e. hydrogen bonds between the N-H (amide) group of the glycines and the C=O (carbonyl) of the *Xaa* residues of neighboring chains: *Gly:NH-OC:Xaa*.

Hydrogen bonds mediated by water molecules present in the second solvation layer only occur in the central region of the peptide, precisely when the residue in the *Xaa* position is not an imino acid - for more details on this interaction, see Ramachandran and Karta (1955). In this case, a second pattern of hydrogen bonds is repeated in the central zone: *Xaa:NH-water-OC:Gly*. The presence of two hydrogen bonds promoted by each tripeptide in this region respects the double hydrogen bond model advocated by Ramachandram and Chandrasekharan (1968).

The low content of imino acids in the central region of T3-785 induces a smooth and localized unfolding of the triple helix, making it more exposed and susceptible to attack by collagenases (FAN et al., 1993). In view of this, a single trypsin cleavage site at position 789-790 (RAMSHAW; SHAH; BRODSKY, 1998) and two sites for *Gly→Ser* and *Gly→Val* mutations are characteristic for *Ehlers Danlos* Syndrome type IV (TROMP et al., 1989; TROMP et al., 1995).

2.4 Technique for a quantum physics study

In this section, the methodological procedures and computational aspects related to the description of the stabilizing forces of the T3-785 peptide will be discussed. The structure of the model was meticulously adjusted and the interaction energies between the residues that make it up were measured using quantum mechanics calculations within the scope of Density Functional Theory (DFT) using the Molecular Fractionation with Conjugated Hoods (MFCC) technique.

2.4.1 Structural data and classical calculations

The crystallographic structure of the T3-T85 peptide was obtained using the X-ray diffraction technique with a resolution of 2.0 Å and stored in the *protein data bank* with the code 1BKV. The computational calculations were carried out on eight XEON servers with 24 cores and 70gb RAM, using *Discovery Studio* and *Materials Studio software*.

Initially, the protein preparation tools in the *Discovery Studio software* were used to generate a summary report on possible structural problems with the T3-785 peptide and, subsequently, corrections were made to the connectivity and charge of the atoms, as well as adding the hydrogens to the system.

Structures determined by crystallography rarely include hydrogens (H), explicitly due to the minimal electronic density of these atoms. In view of this, the free valence of each heavy atom in the system was supplemented with H atoms and, for spatial corrections, geometry optimization calculations were performed. During the optimization of the positions of the hydrogen atoms, all the other atoms in the system were kept fixed.

Initially, the *FORCITE* module, implemented in the *Materials Studio Modeling* software package, was used to run a classic energy minimization procedure, based on the conjugate gradient algorithm, convergence tolerance of 2.0×10^{-5} kcal/mol for the energy variation, 0.001 kcalÂ$^{-1}$ mol^{-1} for the maximum atomic force, 1.0×10^{-5} for the maximum atomic displacement and 14 Â for the cut-off radius of the electrostatic and *van der Waals* interactions. In a second step, quantum geometric optimization processes (DFT) were run from the *DMOL* module (DELLEY, 2000) of the same package, using a threshold for energy variation of 1.0×10^{-5} Ha, maximum force of 0.002 HaÂ$^{-1}$ and maximum displacement of 0.005 Â between subsequent optimization cycles.

2.4.2 Interaction energies and *ab initio* calculations

The interaction energies between the three chains of T3-785 were determined by Quantum Mechanics calculations using Density Functional Theory (DFT) and the Molecular Fractionation with Conjugated Hoods (MFCC) method, which originated from thermochemical theory (GORDON et al., 2012). In this technique, the energy description of a macrosystem can be calculated from the *Schrodinger* Hamiltonian of the parts (fragments) that make it up. In this way, the computational cost is reduced so that precise methods such as DFT can be used. The theoretical principles involved have already been tested and validated, mainly in protein systems (ZHANG; ZHANG, 2003; ZHANG; ZHANG, 2005; MEI et al., 2005; DA COSTA et al., 2012; ZANATTA et al., 2014; MARTINS et al., 2013; RODRIGUES et al., 2013; RIBEIRO, 2014).

Before applying the method, it was necessary to identify which residue-residue interactions would have their energies quantified. To do this, an imaginary sphere with a radius of 8.0 Â was drawn starting from each amino acid of interest (R_i) in order to encompass amino acids (R_j) from other adjacent chains. In this way, we closed interaction sites for the 90 amino acid residues of T3-785, following the order: residues from chain *A* interacting with chain *B* (subsystem *AB*); residues from *B* interacting with *C* (subsystem *BC*); and residues from *C* with those located in *A* (subsystem *CA*).

After this stage, the system was fractionated using the MFCC method. Four fragments were formed for each R_i. The first, $C_{i-1}R_iC_{i+1}-C_{j-1}R_jC_{j+1}$, is made up of the residues R_i and R_j, together with those that precede/succeed them, in this case C_{i-1}/C_{i+1} and C_{j-1}/C_{j+1}, respectively. Subsequently, only R_i and only R_j were removed to obtain the second, $C_{i-1}R_iC_{i+1}-C_{j-1}C_{j+1}$, and the third, $C_{i-1}C_{+1}-C_{j-1}R_jC_{j+1}$, fragments, respectively. Using the same reasoning, the fourth fragment, $C_{i-1}C_{+1}-C_{j-1}C_{j+1}$, was constructed by excluding R_i and R_j at the same time.

The underlying and adjacent residues are called "hoods" and are present in order to reliably represent the chemical and electronic environment surrounding R_i and R_j (CHEN; ZHANG, 2004). To this end, each water molecule was attached, as an integral part, to the nearest residue with which it was hydrogen bonding. If the water molecule was not hydrogen bonding, it was attached to the nearest residue. Thus, the effects of microsolvation were taken into account when calculating the energy bond between R_i and R_j.

The residue-residue interaction energy, E_I (R_I - R_j), was calculated using the equation:

$$E_I (R_i - R_j) = E(C_{i-1}R_iC_{i+1}-C_{j-1}R_jC_{j+1}) - E(C_{i-1}R_iC_{i+1}-C_{j-1}C_{j+1}) - E(C_{i-1}C_{i+1}-C_{j-1}R_jC_{j+1}) + E(C_{i-1}C_{i+1}-C_{j-1}C_{j+1}) \ (1)$$

$E(C_{i-1}R_iC_{i+1}-C_{j-1}R_jC_{j+1})$ is the total energy of the fragment formed by the two mutually interacting residues and their hoods; the term $E(C_{i-I}R_I C_{i+I}-C_j-_{1C_{j+1}})$ corresponds to the energy of the residue R_i with its $C_{i-1}C_{i+1}$ hoods, together with the energy of the $C_{j-1}C_{j+1}$ hoods of R_j ; $E(C_{i-1}C_{i+1}-C_{j-1}R_jC_{j+1})$ is the energy of the system formed by R_j and the other hoods; finally, $E(C_{i-}1C_{i+1}-C_{j-1}C_{j+1})$ indicates the energy of the fragment formed only by the R_i and R_j hoods. A schematic representation of this process can be seen in Figure 9.

In this work, the energy calculations were carried out through quantum simulations executed within the formalism of density functional theory (DFT), using the *Perdew-Wang* local density approximation (LDA-PWC) and the *Perdew-Burke-Ernzerhof* generalized gradient approximation (GGA-PBE) for the determination of the exchange-correlation potential (PERDEW; BURKE; ERNZERHOF, 1996) together with a double precision plus polarization (DNP) basis for expansion of the *Kohn-*

Sham wave functions, taking all electrons into account explicitly with unrestricted spin. Deley (1999) and Inada and Orita (2008) point to the high accuracy and practically irrelevant basis set superposition error (BSSE) of this basis set. The DNP is comparable to the Gaussian basis set 6-311+G(3df,2pd), which has a large set of polarization/diffuse functions and an effective electronic correlation (INADA; ORITA, 2008; DELLEY, 1990).

Figure 9 - Molecular Fractionation Method with Conjugated Hoods (MFCC) exemplified by the energy calculation of the residue-residue interaction between Pro7 (chain A) and Hyp38 (chain B).

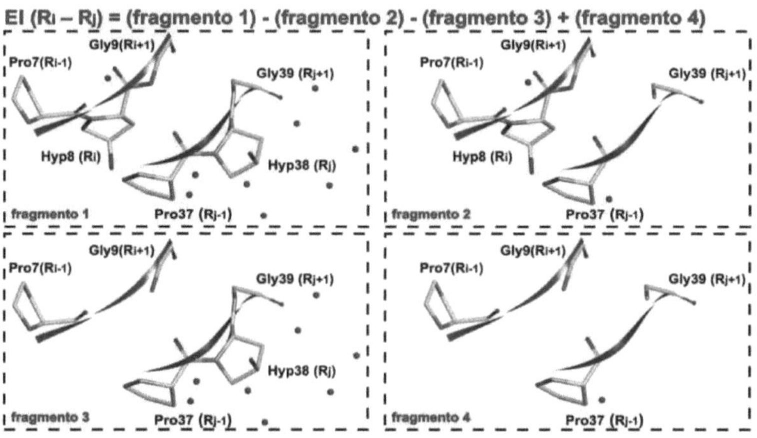

Source: Prepared by the author

To guarantee the quality of the description of the atomic energies, the program was set to *Fine* precision during the numerical integration of the system's Hamiltonian, with an orbital cutoff radius of 5.5 Å and a self-consistent field (SCF) convergence threshold set for an energy variation of up to 10^{-6} Ha.

It is well established that the addition of dispersion terms in DFT calculations is extremely important in the correct qualitative description of the non-covalent bonds in a system (ANTONY et al., 2011; GRIMME, 2012). Antony and Grimme (2012) demonstrated that the MFCC technique, in conjunction with these terms, adequately dealt with aspects such as exchange-repulsion, electrostatics, induction and dispersion in a protein-ligand complex. In this way, it was accurate and efficient in modeling the system's intermolecular interactions.

A recent study evaluated the effectiveness of the Grimme (GRIMME, 2006), Ortmann-Bechstedt-Schmidt (ORTMANN; BECHSTEDT; SCHMIDT, 2006), and Tkatchenko-Scheffler (TS) (TKATCHENKO; SCHEFFLER, 2009) dispersion corrections during energy and electronic property calculations of the porphyrin-fluorene system, which is rich in non-covalent interactions. Basiuk and Henao-Holg (2014) observed that the results obtained with GGA/PBE/Grimme and the OBS correction provide a better description of hydrogen bridges (*H-Bond*) and *van der Waals* interactions (*vdW*) in calculations of various properties based on density functional theory (BASIUK; HENAO-HOLG, 2014).

In order to adequately describe the intermolecular forces between the amino acid residues of the T3- 785 triple helix, the OBS and Grimme methods were used in the LDA-PWC and GGA-PBE calculations, respectively. In this way, six terms are contained when the interaction energy between two residues of interest located in adjacent chains of T3-785 collagen is described: (1) proton-proton repulsion + (2) electron-proton attraction + (3) electron-electron coulombic repulsion of a classical nature + (4) electron kinetic energy + (5) non-classical electron-electron exchange energy + (5) correlation energy, which describes the correlated movement of electrons with opposite spins.

In order to avoid too many numerical figures, the data presented in this book is restricted to that generated by LDA-PWC-OBS. However, it is important to clarify that absolutely no conclusions reached here are contrary to the data generated via GGA-PBE-Grimme.

2.4.3 Intermolecular contacts

The computational calculations were designed to ensure an adequate description of the non-covalent interactions at the interface of the T3-785 chains, be they hydrogen bonding, ion-dipole, induced ion-dipole or London dispersion forces.

The signature of the system's hydrogen bonds was based on angular criteria and the distance between the donor and acceptor atoms. The hydrogen-acceptor distance ≤3.1 Å and the donor-hydrogen-acceptor angle between 90° and 180° were considered. These values encompass 90% of the bonds statistically observed in structures deposited in the *Cambridge Structural Database* (CSD) (BISSANTZ;

KUHN; STAHL, 2010). In biological systems, an unconventional hydrogen bond is possible between a donor of low electronegativity, commonly a carbon atom, and a traditional receptor of high electronegativity, such as nitrogen or oxygen (WAHL; SUNDARALINGAM, 1997; MANIKANDAN; RAMAKUMAR, 2004; CHAKRABARTI; BHATTACHARYYA, 2007). Derewenda, Lee and Derewenda (1995) characterized a large proportion of unconventional hydrogen bonds by studying very precise crystallographic data of 13 proteins. In these structures, C^{α} H units acted as weak donors in unconventional C^{α} H-OC hydrogen bonds, at an average distance of 3.5 Â between C^{α} and O.

The other intermolecular contacts were considered to be close to those observed for conventional hydrogen bonds.

2.5 Structural and energetic characterization of collagen

The structural stability of the human collagen triple helix is commonly studied from a reductionist perspective, whereby model peptides with high physical similarity have their conformational properties analyzed and subsequent conclusions correlated (BELLA et al., 1994; BELLA; BRODSKY; BERMAN, 1995; KRAMER et al., 1998; KRAMER et al., 1999; KRAMER et al., 2000; KRAMER et al., 2001; BERISIO et al..., 2000; VITAGLIANO et al. 2001; BERISIO et al., 2002; OKUYAMA et al., 1981; NAGARAJAN; KAMITORI; OKUYAMA, 1999; OKUYAMA et al., 2004; JIRAVANICHANUN et al., 2005; JIRAVANICHANUN; NISHINO; OKUYAMA, 2006; OKUYAMA et al, 2007; EMSLEY, 2000; SCHUMACHER; MIZUNO; BACHINGER, 2006; SCHUMACHER; MIZUNO; BACHINGER, 2005; KAWAHARA ET AL., 2005; HONGO ET AL., 2001; OKUYAMA et al., 2004; HONGO et al., 2005; OKUYAMA et al., 2011; OKUYAMA et al., 2012). In this perspective, we used crystallographic data of the tropocollagen T3-785 (PDB ID: 1BKV), obtained by X-ray diffraction technique, as a target model for a chemical analysis of the stabilizing forces of its conformational state.

The T3-785 molecule is in the homotrimeric triple helix state, where each chain is made up of three repeats of *Pro-Hyp-Gly* (N-terminal region or zone), followed by the sequence *Ile-Thr-Gly-Ala-Arg-Gly-Leu-Ala-Gly* (central region or zone) and then four more repeats of *Pro-Hyp-Gly* (C-terminal region). Figure 5 shows this structure obtained by PDB 1BKV, with three intertwined monomers, designated by the letters A (blue), B (green) and C (yellow), in an aqueous medium. It can be seen that chains *A*, *B* and *C* comprise residues 1-30, 31-60 and 61-90, respectively. The optimal structure of T3-785 achieves helical stability and exhibits many of the general characteristics observed in other collagen models (SCHUMACHER; MIZUNO; BACHINGER, 2005).

This structure is sequentially analyzed for its energetic stability. Initially, quantum calculations (see details in the *Methodology* section) are carried out to quantify the residue-residue (i.e. pairwise) interaction energies of the amino acids that make up T3- 785, in the following order: residues of the *A* chain interacting with residues of the *B chain* (together they form the *AB* subsystem), residues of the *B chain* interacting with residues of the *C chain* (*BC subsystem*), residues of *C* with *A* (*CA subsystem)* - Figure 10. The sum of these energies is referred to during the analysis of the entire system, called the *ABC* complex.

Secondly, the energetic attractions of the N-terminal, central and C-terminal zones were calculated, as well as those of the *Pro-Hyp-Gly, Ile-Thr-Gly, Ala-Arg-Gly* and *Leu-Ala-Gly* triads that make up the *A*, *B* and *C* chains in the *AB*, *BC* and *CA* subsystems, respectively. Finally, we analyzed the individual relevance (importance) of the amino acids *Pro*, *Hyp*, *Gly*, *Ile*, *Thr*, *Ala*, *Arg* and *Leu* for the stabilization of the triple helix model. Finally, we evaluated the energetic impact in terms of stability of point substitutes on the wild-type structure of collagen.

It is important to note that the positively (negatively) charged N-(C-)terminal residues *Pro31*, *Pro61* and *Hyp2* (*Gly30, Gly60 and Gly90)* of the *A*, *B* and *C* chain, respectively, and consequently, the sequences *Hyp2-Gly3*, *Pro31-Hyp32-Gly33* and *Pro61-Hyp62-Gly63* (*Pro28-Hyp29-Gly30*, *Pro58-Hyp59-Gly60* and *Pro88-Hyp89-Gly90*) were disregarded during all the energetic analyses of the triple helix. In fact, the *Pro- Hyp-Gly* triads at the ends of T3-785 have a low crystallographic quality, as evidenced by the poor local electronic density and the atoms signed with high temperature factors (KRAMER et al., 2001) - Figure 11. The atomic positions in these regions are highly inaccurate, so they were disregarded, as previous studies have done (KRAMER et al., 1999; KRAMER et al., 2001).

Figure 10 - Identification of the *AB, BC and CA* subsystems on the T3-785.

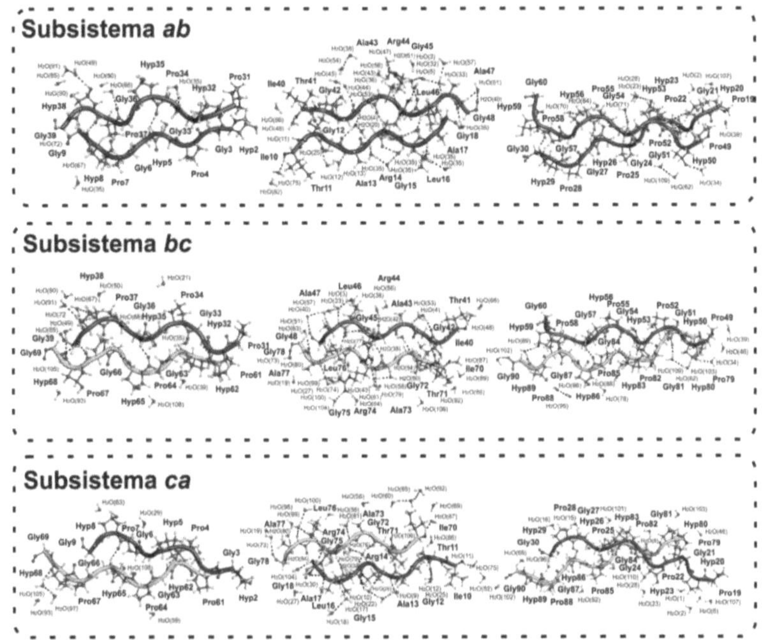

Source: *Prepared by the author*

Figure 11 - Representation of T3-785 according to the temperature factors (*B-factor*) signed for each atom. The color variation from blue to red represents the *B-factor* range of 10-100 Å2.

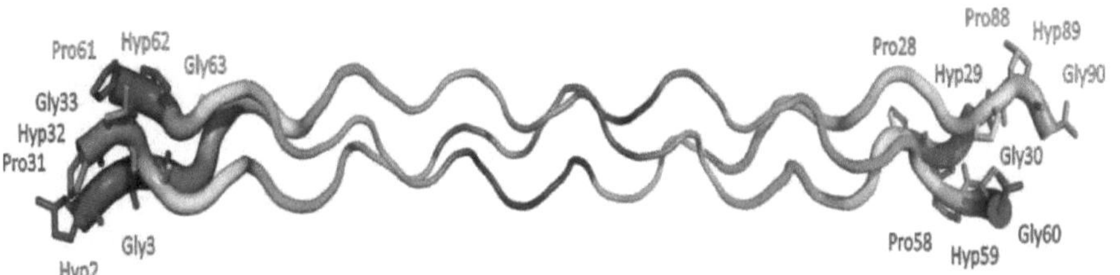

Source: *Prepared by the author*

2.5.1 Energy aspects of the zones

Table 2 shows that chain *A* (*B*) interacts with chain *B* (*C*) with a binding energy of - 625.67 kcal/mol (-566.12 kcal/mol). The total interaction of the *C* chain with the *A chain* has an intensity of - 566.31 kcal/mol. Thus, the T3-785 triple helix (*ABC* system) has a quantified stability of -1758.10 kcal/mol. The central region accounts

37

for 42.09% of the triad's attraction energy, while the C-terminal and N-terminal zones account for 25.29% and 32.61%, respectively. The fact that the latter is made up of four *Pro-Hyp-Gly triads,* while the other is made up of only three, explains the difference of approximately 8% between them.

Individually, we can see that for *AB* and *CA,* the order of importance N-terminal < C-terminal < Central of the *ABC* system is respected. In *BC,* the influence of the last two regions of the triple helix is reversed. This is due to the energy of attraction of -187.39 kcal/mol relating to the interaction of the residues of the central zone of chain *B* with those of *C,* which is considerably lower than the values of -252.51 kcal/mol and -300.13 kcal/mol of the *CA* and *AB* systems, respectively. This difference will be explained below by the difference in the interactions between the residues of argenines from different chains.

Table 2 - Binding energies of the N-terminal, central and C-terminal zones that make up each of the A, B and C chains in interaction with the B, C and A chains, respectively. In particular, the central zone of the ABC system is responsible for 42% of the total force of attraction of the collagen triple helix. All values are in the unit kcal/mol.

	AB system		BC system		AC system		ABC System	
N-Terminal*	-129,77	20,74%	-171,86	30,36%	-143,05	25,26%	-444,69	25,29%
Central Zone	-300,13	47,97%	-187,39	33,10%	-252,51	44,59%	-740,03	42,09%
C-Terminal*	-195,77	31,29%	-206,87	36,54%	-170,75	30,15%	-573,38	32,61%
TOTAL	-625,67	100,00%	-566,12	100,00%	-566,31	100,00%	-1758,10	100,00%

The interaction energies related to the terminal residues Pro31/61, Hyp2/32/62 and Gly33/63 were disregarded.

Source: *Prepared by the author*

2.5.2 Energy aspects of triads

As the structure of collagen is made up of triads that are repeated throughout its structure, the importance of these for the model peptide was assessed comparatively. Table 3 shows the average interaction energies of the triads that make up chains *A*, *B* and *C* with chains *B*, *C* and *A,* respectively. It's important to note that the aim of this table is to show the stability promoted by the mere presence of each of the triads, regardless of the number of times they are repeated in the model, which is why we used the average interaction values. So, for example, the value of -65.11 kcal/mol for *Pro-Hyp-Gly* corresponds to the average value of all seven *Pro-Hyp-Gly* sequences in T3-785.

In terms of attractiveness, the triads *Ala13-Arg14-Gly15 (A), Ala43- Arg44-Gly45*

(B) and *Ala73-Arg74-Gly75 (C)* stand out from the rest, as they account for a total binding energy of -130.15 kcal/mol. An interaction energy almost twice as high as any other. The *Pro-Hyp-Gly, Ile-Thr-Gly and Leu-Ala-Gly* triads, present in the three α-chains of T3-785, interact with very close average energies (-67.87 kcal/mol, -65.23 kcal/mol and -55.12 kcal/mol).

Table 3 - Average bond energies of the A, B and C chain triads in interaction with the B, C and A chains, respectively. Particularly noteworthy is the Ala-Arg-Gly triad, which has a total E.I. approximately twice as high as the others. All values are in the unit kcal/mol.

	AB system		BC System		AC system		ABC System	
Pro-Hyp-Gly*	-65,11	17,27%	-75,74	28,79%	-62,76	19,91%	-67,87	21,31%
Ile-Thr-Gly	-74,64	19,80%	-61,90	23,53%	-59,15	18,76%	-65,23	20,48%
Ala-Arg-Gly	-167,57	44,45%	-68,06	25,87%	-154,83	49,11%	-130,15	40,87%
Leu-Ala-Gly	-69,64	18,47%	-57,42	21,82%	-38,53	12,22%	-55,20	17,33%
TOTAL	-376,96	100,00%	-263,12	100,00%	-315,27	100,00%	-318,451	100,00%

The interaction energies of the triads with the terminal residues Pro31⁄61, Hyp2⁄32⁄62 and Gly33⁄63 were disregarded.

Source: *Prepared by the author*

2.5.3 Energy description of waste

Table 4 shows the energetic link between the residues that make up *A*, *B* and *C and* the *B*, *C* and *A chains,* respectively. The energy values shown reflect the affinity between the chains of the *AB*, *BC* and *CA* subsystems, based on the average attraction promoted by each type of amino acid residue present in *A, B and C,* respectively. By way of example, the average attraction of alanine residues in the *AB subsystem* (called "residue-subsystem" energy, in this case, "*Ala-AB*") is formed by the average energy of all the interactions of *Ala13* and *Ala17* with residues of the *B* chain within a radius of 8 Â. In this case, the interactions *Ala13-Ile40, Ala13-Thr41, Ala13-Gly42, Ala13-AlaA43, Ala13- ArgG44* and *Ala13-GlyY45* were calculated and added together, in addition to *Ala17-ArgG44, Ala17-Gly45, Ala17-Leu46, Ala17-AlaA47, Ala17-GlyLY48* and *Ala17-Pro49* (each of these is called "residue-residue" energy), then the arithmetic mean was found.

Table 4 - Average binding energies of amino acid residues from chains A, B and C in contact with chain B, C and A, respectively. The order of relevance is Arg > Hyp > Thr > Ala > Gly > Pro > Leu > Ile. All values are in the unit kcal/mol.

	AB system		BC System		AC system		ABC System	
Pro*	-7,71	2,88%	-10,95	5,77%	-5,94	2,42%	-8,20	3,50%
Hyp*	-38,66	14,42%	-48,30	25,46%	-36,75	15,00%	-41,24	17,60%
Gly*	-18,99	7,08%	-17,02	8,97%	-18,81	7,67%	-18,27	7,80%
Ile	-6,10	2,28%	0,40	-0,21%	-7,45	3,04%	-4,38	1,87%
Thr	-40,05	14,94%	-48,07	25,34%	-27,39	11,18%	-38,50	16,43%

Wing	-39,09	14,58%	-20,29	10,69%	-18,84	7,69%	-26,07	11,13%
Arg	-110,23	41,11%	-42,46	22,39%	-124,94	50,98%	-92,54	39,50%
Read	-7,31	2,73%	-2,99	1,58%	-4,95	2,02%	-5,08	2,17%
TOTAL	-268,14	100,00%	-189,68	100,00%	-245,06	100,00%	-234,30	100,00%

The interaction energies of the triads with the terminal residues Pro31/61, Hyp2/32/62 and Gly33/63 were disregarded.

Source: *Prepared by the author*

It should be noted that, of course, the energy added up in each column does not represent the total energy of the subsystems, but rather the sum of the average energy values of each type of residue - in this case *Pro, Hyp, Gly, Ile, Thr, Ala, Arg* and *Leu*. The way in which the interaction energies have been displayed in Table 4 provides a reflection on the individual relevance (contribution) of each T3-785 residue to the stabilization of the system. Thus, in this book, the quality of a given residue has not been given by the number of times it appears in the structure, but by the strength with which it acts individually in the system.

In fact, a careful analysis of Table 4 shows that there is an order of relevance among the residues that make up T3-785: *Arg* (with a relative influence of 39.5%) > *Hyp* (17.60%) > *Thr* (16.43%) > *Ala* (11.13%) > *Gly* (7.80%) > *Pro* (3.50%) > *Leu* (2.17%) > *Ile* (1.87%). Depending on position, *Ala* > *Pro* > *Leu* > *Ile* (*Arg* > *Hyp* > *Thr* > *Ala*) were observed for those located in *Xaa* (*Yaa*). At this point, the data obtained differs partially from that found by Persikov et al. (2000), namely: *Pro* > *Ala* > *Leu* > *Ile* (*Hyp* > *Arg* > *Ala* > *Thr*). These differences could be explained if we assume that the molecular interactions of each residue in a triad are influenced by its adjacent elements. This was in fact observed by Persikov and his team in their study entitled "Amino Acid Propensities for the Collagen Triple-Helix", in which they used the central triads of the (*Gly- Pro - Hyp*)3-Gly- *Xaa-Hyp-(Gly-Pro-Hyp)*4 and *(Gly-Pro-Hyp)3-Gly-Pro-Yaa-(Gly-Pro-Hyp)*4 models to demonstrate the propensity/tendency of the 20 types of amino acids to occupy the *Xaa* and *Yaa* positions, taking into account differences in denaturation temperatures. In this study, the order of importance of a given amino acid varied depending on the substitution site, *Gly-Xaa - Hyp* or *Pro - Yaa-Gly*. Therefore, adjacent residues, necessarily *Gly/Hyp* or *Pro/Gly*, influence the thermostability of the central element.

The order of relevance shown in this work gives indications of the molecular interactions that govern T3-785. The most important residues - *Arg, Hyp* and *Thr* - have side chains capable of influencing nearby residues through molecular interactions of a fundamentally electrostatic nature. In the model used, it was observed that the guanidine group [RNHC(NH2)2+] of *Arg* can make charge-dipole, charge-induced dipole and dipole-dipole interactions, while the hydroxyl (OH) of *Hyp* and *Thr* commonly makes hydrogen bonds (dipole-dipole) or induced dipole-dipole. The other residues interact dispersively (*London* dispersion forces) because they have aliphatic side chains capable of undergoing instantaneous fluctuations (femtosecond order) in their electronic densities. Obviously, every amino acid can make hydrogen bonds from its main chain (ketone or amine group).

The importance of intermolecular interactions for the structural preservation of triple helix systems was highlighted by Hodges and Raines (2005) when they obtained heterotrimeric triple helices from the association of individual peptides incapable of forming stable helices.

Electrostatic interactions strongly influence the stability of collagen models. The development of collagen-like peptides formed by stable heterotrimeric helices represents one of the most solid examples of triple helix stabilization through electrostatic interactions (GAUBA; HARTGERIN, 2007a; GAUBA; HARTGERIN, 2007b; FALLAS; GAUBA; HARTGERINK, 2009). Experimental data on fusion temperature (Tm) and *Gibbs* free energy indicated strong favorable (unfavorable) electrostatic interactions for the *Gly-Lys-Asp* and *Gly-Arg-Asp* triads (*Gly-Arg-Lys*, *Gly-Lys-Arg*, and *Gly-Glu-Asp*) when present in the *acetyl-(Gly-Pro-Hyp)3-Gly-Xxx-Yyy-(Gly-Pro- Hyp)4-Gly-Gly-NH2* model (PERSIKOV et al., 2002).

Argeninas

Systems with the *Gly-Xaa-Yaa* triad in the central region of each chain have gradually decreasing thermal stability as the number of *Arg* in the *Yaa* position increases when *Pro* permanently occupies the *Xaa* position (KOIDE; NISHIKAWA; TAKAHARA, 2004). However, in the presence of *Arg* and *Glu* in *Xaa* and *Yaa*, respectively, strong interchain electrostatic interactions increase the stability of the heterotrimer triple helix (GAUBA; HARTGERINK, 2007a). In the T3-785 peptide, the three argenines of the Arg14|44|74 system occupy the Yaa position of their respective triads. Conformational, CD *spectra* and thermal stability data have indicated that *Arg* residues are largely favorable when they occupy the *Yaa* position

(BERISIO et al., 2009; YANG et al., 1997; PERSIKOV et al., 2000).

Figure 12 shows the energies involving the *Arg* residues in the *A*, *B* and *C* chains and their closest neighbors belonging to the *B*, *C* and *A* chain, respectively. *Arg14* (*Arg44*) engages attractively with residues *Ala43*, *Gly45*, *Leu46* and *Ala47* (*Ala73*, *Gly75*, *Leu76* and *Ala77*) with energies of -42.57 kcal/mol, -8.50 kcal/mol, -54.76 kcal/mol and -12.28 kcal/mol (-33.32 kcal/mol, -4.65 kcal/mol, -11.35 kcal/mol and -2.54 kcal/mol), respectively. The *Arg74* residue of the *C* chain exhibits four interactions of the same nature with the residues of the *A* chain, namely: *Gly15* (-1.45 kcal/mol), *Leu16* (-50.47 kcal/mol), *Ala17* (-38.27 kcal/mol), *Gly18* (-7.85 kcal/mol), *Pro19* (-41.63 kcal/mol) and *Hyp20* (-12.98 kcal/mol).

Figure 12 - Binding energies of amino acid residues interacting with Arg14, Arg44 and Arg74 in the *AB*, *BC* and *CA* subsystems, respectively.

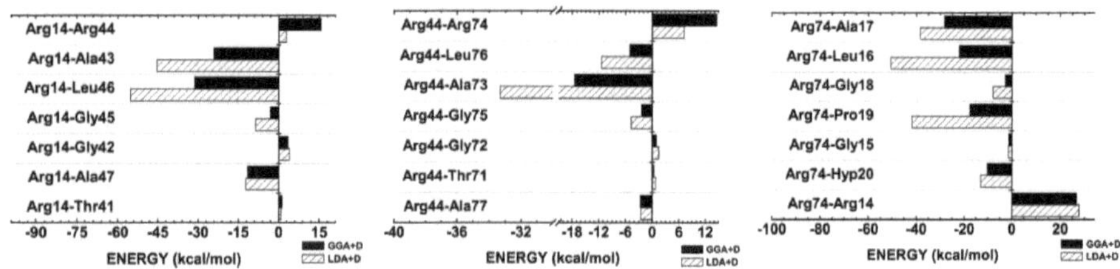

Source: *Prepared by the author.*

Table 4 shows that *Arg* is the residue with the greatest individual relevance for the conformational stability of T3-785, accounting for 39.50% of the average attractiveness of the *ABC* system. Individually, *Arg14|44|74* residue-residue interactions account for 41.11%|23.39%|50.98% of the affinity in *AB*, *BC* and *CA*, respectively. Although it is still one of the main residues in *BC*, *Arg44*'s interaction energy of -42.46 kcal/mol is much lower than that observed for *Arg14* (-110.23 kcal/mol) and *Arg74* (-124.96 kcal/mol) in their respective subsystems. *Arg44* has only one strong residue-residue interaction, with *Ala73*, while *Arg14* is highly attracted to *Ala43* and *Leu46*, and *Arg74* to *Leu16*, *Ala17* and *Pro19*. In order to understand the difference in these energy profiles, structural and solvation aspects of each residue involved were analyzed. The argenines differ in the amount of water molecules able to interact with their lateral functional groups and other amino acid residues nearby (Figure 13). Interestingly, it was found that the (average) interaction energy of *Arg14-AB, Arg44-BC* and *Arg74-CA* is proportional to the number of water molecules present, namely five, four and seven, respectively.

Figure 13 - Most effective energetic interactions of residues Arg14, Arg44 and Arg74. The interactions of Arg14 (Arg44) with Leu46 (Leu76) and Ala43 (Ala73) occur with the participation of induced charge-dipole contacts and bifurcated hydrogen bonds respectively. Arg44-Leu76 has a lower energy than Arg14-Leu46 due to the greater distance between its side chains. The Arg74-Leu16, Arg74-Pro19 and Arg74-Ala17 interactions stand out for their two induced charge-dipole contacts and one hydrogen bond respectively.

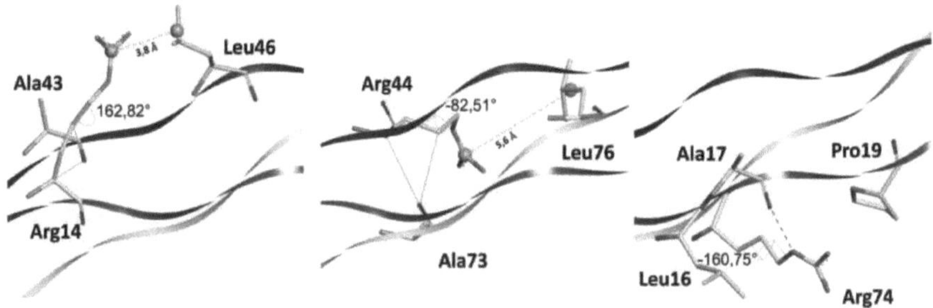

Source: *Prepared by the author.*

In conformational terms, it was found that the dihedral angles χ3 (C -C -C$^{\beta\gamma\delta\epsilon}$) - Nequal to $162.82°$ | $-160.75°$ of the *Arg14174* residues determine a *trans* conformational state for their side chains, which favors the approximation between the guanidine groups and the *Leu46116* residues and, consequently, the strengthening of the induced charge-dipole interactions. The calculated energies of -42.57 kcal/mol and -38.17 kcal/mol are justified by the existence of a bifurcated non-conventional hydrogen bond between *Arg14* and *Ala43* (Arg14O$^{\alpha}$ H | C γ H– OC:*Ala43*) and a conventional one between the *Arg74-Ala17* pair (*Arg74*:NsH-OC:*Ala17*). In the first case, a single oxygen atom can simultaneously participate as a receptor in two hydrogen bonds (ROZAS; ALKORTA; ELGUERO, 1988; PADIYAR; SESHADRI, 1996; ISAEV, 2013). Additionally, we believe that the high affinity of the charge-dipole interaction induced between *Arg74* and *Pro19* is also favored by the *trans* conformation that ensures parallelism between their side chains.

Unlike *Arg14* and *Arg74*, the side chain of *Arg44* acquires a *gauche* conformation·

due to its χ3 angle of - 82.51°. In this state, there is a slight displacement of the guanidine group in the opposite direction to the position of the residues with which it interacts, with the exception of *Ala73*. In fact, the distance between the centroids of the guanidine group of *Arg44* and the isobutyl of *Leu76* is approximately 5.6 Å, while in the *Arg14-Leu46* system it does not reach 3.8 Å. With regard to *Arg44-Ala73*, there is a hydrogen bond (*Arg44* O^α H | C γ H–OC: *Ala73*) very similar to that observed in *Arg14 - Ala43,* which translates into very close energetic bonds.

The high repulsiveness of the *Arg74-Arg14, Arg44- Arg74 and Arg14-Arg44* pairs is related to the strong charge-charge interaction between the guanidine functional groups of the same charge.

Proline and hydroxyproline

The X and Y positions are often occupied by the amino acids proline (Pro, 28.1%) and hydroxyproline (Hyp, 38.1%), respectively. Together they account for approximately 22% of all collagen ribbon residues (RAMSHAW; SHAH; BRODSKY, 1998).

It is well known that the bulky and inflexible lateral chains of *Pro* and *Hyp* provide a strong rigidity capable of supporting the spatial positioning of the triple helix structure. In the T3-785 peptide, the *Pro* and *Hyp* residues occupy the *Xaa* and *Yaa* positions respectively in all the *Gly- Xaa-Yaa* triads of the terminal regions. Considering the radius of 8 Å from each *Pro*, the residues *Pro417119122125, Pro34137149152155* and *Pro64167179182185* interact with several residues of the *B,CeA* chain, respectively, totaling an energy of -6.58|-1.50|-18.39|-6.26|-5.82 kcal/mol, -18.66|- 4.78|-6.89|-16.67|-7.75 kcal/mol and -3.33|-5.32|-7.64|-5.67|-7.73 kcal/mol for each of them, respectively. Thus, the (average) *Pro-AB, Pro-BC* and *Pro-CA* energies are equal to -7.71 kcal/mol (2.88% of the average *AB* subsystem interaction), -10.95 kcal/mol (5.77%) and -5.96 kcal/mol (2.42%), respectively. With regard to residues *Hyp518120123126* from *A* interacting with *B* (subsystem *AB*), *Hyp35|38|50|53|56* from *B* interacting with *C* (subsystem *BC*) and *Hyp65|68|80|83|86* from *C* interacting with *A* (subsystem *CA*), the following corresponding energy values were quantified: -31.13|-45.25|-60.68|-29.22|-27.05 kcal/mol, -27.17|-85.63|- 44.69|-43.32|-40.69 kcal/mol and -32.57|-52.65|-55.36|-

21.11|-22.06 kcal/mol. Consequently, average energies of -38.66 kcal/mol (14.42% of *Hyp-AB energy*), -48.30 kcal/mol (25.46% of *Hyp-BC energy*) and -36.76 kcal/mol (15% of *Hyp-CA energy*).

In 1988, Bhatnagar et al. proposed that *van de Walls* interactions of the *Pro : Pro* type are important for stabilizing triple helix structures. However, of the 35 *Pro-Pro* interactions evaluated here, few showed any degree of attractiveness. The pyrrolidine rings of the interchain prolines are relatively far apart ($dist_{C\gamma-C\gamma}$ > 7.9 Å), which makes strong Pro:*Pro van der Walls* interactions impossible, as observed in (*Pro-Pro-Gly*)9 crystals, in which prolines from adjacent chains have $dist_{C\gamma\text{-}C\gamma}$"4.1A (HONGO et al., 2005) - see Figure 14.

All hydroxyprolines are highly attractive to some of the proline residues in parallel chains due to unconventional hydrogen bonds - see details *in* the topic *Hydrogen Bonding Patterns*.

Figure 14 - Inability of hydrophobic contacts involving proline molecules.

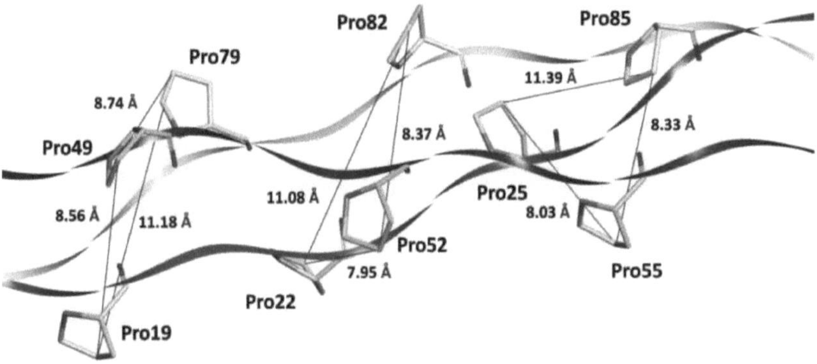

Source: *Prepared by the author.*

Our calculations show that the greatest residue-residue attractions involving the amino acids proline and hydroxyproline occur in the presence of unconventional hydrogen bonds. In fact, the carbonyls of the prolines located in the *A|B|C* chains make contacts of this nature with the C^{α} H of the glycines in *B|C|A*, respectively: *Pro4|34|64- Gly33|63|6, Pro7|37|67-GlyY36|65|9, Pro19|49|79-Gly48|78|21, Pro22|52|82- Gly51181124, Pro25155185- Gly54184127*. In the case of the *Hyp* residues, we observed that all of them interact with proline from nearby chains, not through the carbonyl oxygen, but through the hydroxyl: *Hyp5|35|65- Pro34|64|4, Hyp8|38|68-Pro37|67|7, Hyp20|50|80-Pro49|79|19, Hyp23|53|83- Pro52|82|22 and Hyp26|56|86-Pro55|85|25.*

Threonines

The *Thr11* residue is responsible for 14.94% (-40.05 kcal/mol) of the attractiveness of the *AB* subsystem, given the sum of the average energy of each type of amino acid in *A* interacting with *B*. *Thr41* (*Thr71*) has a binding energy of -48.07 kcal/mol (-27.39 kcal/mol) for *Thr41-AB* (*Thr71-BC*), which corresponds to 25.34% (11.18%) of the average attraction of the *B* (*C*) chain to *C* (*A*). Considering the entire triple helix, the total (average) energy of attraction is -38.50 kcal/mol (16.43%). Therefore, the *Thr* amino acid is the third most important for maintaining the conformational integrity of the T3-785 peptide. Among all the *Thr-residue* interactions, the two that stand out the most are those in which water molecules mediate the hydrogen bond between the polar group of threonines 11 and 41 and the *N* atom of *Ile40* and *Ile70*, respectively - Figure 15 and 16. Namely: *Thr11-Ile40* (-31.39 kcal/mol), *Thr41-Ile70* (-43.86 kcal/mol). A third interaction, *Thr71-Ala13* (-28.77 kcal/mol), is characterized by an unconventional hydrogen bond between C^{α} H from *Thr71 and CO* from

Ward 13.

Figura 15 - **Interchain bond energies of the aliphatic residues (a) Ile and (b) Leu, considering a radius of 8 Å and the directions A->B, B->C and C->A.**

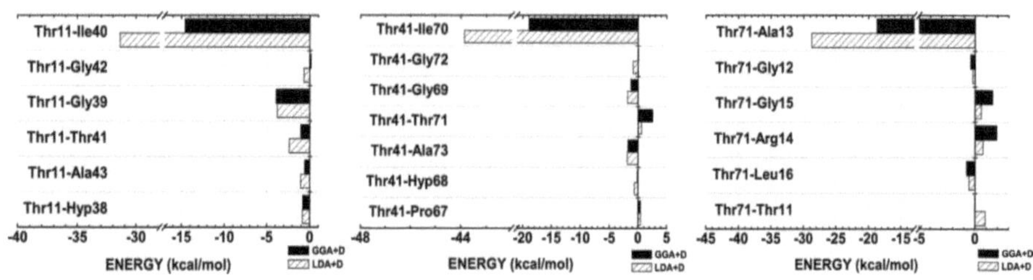

Source: *Prepared by the author.*

Figura 16 - **Main intermolecular contacts of Thr11, Thr41 and Thr71 in the *AB*, *BC* and *CA* subsystems, respectively.**

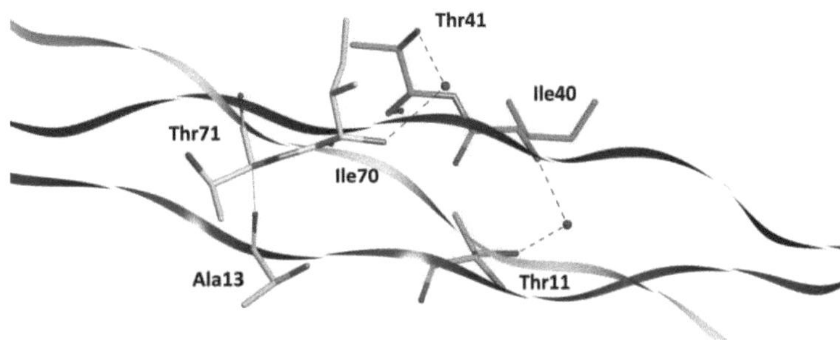

Source: *Prepared by the author.*

Alanines

Unlike the other residues, alanine is present in both positions of the *Gly-Xaa-Yaa* triad, in which case *Ala13117*, *Ala43147*, and *Ala73177* occupy the *Xaa|Yaa* positions on the *A,BeC* chains, respectively. When located in *Xaa* and *Yaa*, alanine interacts with average energies of -14.56 kcal/mol and -37.58 kcal/mol, respectively. These residues are not conformationally distinct from each other - *Ala13*, *Ala43* and *Ala73* (*Ala17*, *Ala47* and *Ala77*) have *phi/psi* dihedral angles of - 75/159, -78/156 and -65/163 (-61/149, -66/152 and -68/152). However, the distances between *Ala(Xaa)*-residue interchain interactions are on average 28% greater than those observed in *Ala* (*Yaa*)-residue.

The greater individual relevance of Alanine in occupying the *Yaa* position is justified by the high attractiveness of *Ala17*, *Ala47* and *Ala77* for residues located in the *Xaa* position of adjacent chains, namely: *Ala17-Leu46* (23.05), *Ala17-Pro49* (-11.68), *Ala47-Leu76* (-13.76), *Ala47-Pro79* (-16.39) and *Ala77-Pro19* (-11.74) - Figures 17 and 18. Due to the chemical nature of the amino acids involved, these contacts probably have a strong hydrophobic character (dispersive interactions). In addition, Ala17-Leu46 and Ala47-Leu76 also make unconventional hydrogen bonds.

Figura 17 - Binding energies of the amino acid residues that interact with the allenin residues located in the (a) Xaa and (b)Yaa positions of T3-785.

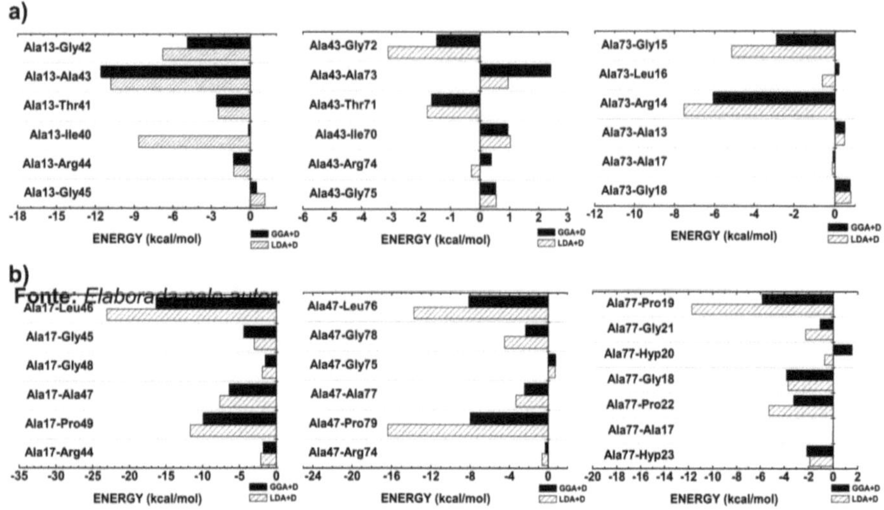

Figura 18 - Main intermolecular interactions promoted by the alanine residues located in the Yaa position (Ala17 | 47 | 77) of T3-785.

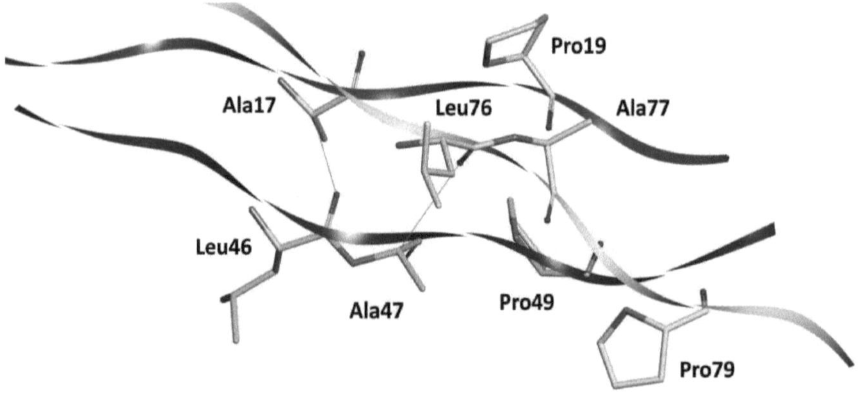

Source: *Prepared by the author.*

Leucine and Isoleucine

According to a renowned study which evaluated the distance of amino acids in the *Gly-Xaa-Yaa* triads in collagens, the *Ile* and *Leu* residues which differ in the isomerism of the (*sec* or *iso*)butyl radical occupy the *Xaa* position in 11.2% of the situations (RAMSHAW; SHAH; BRODSKY, 1998). This homogeneity justifies the proximity of the values of -4.38 kcal/mol and -5.08 kcal/mol for the average interaction energies obtained for these residues.

We believe that the importance of residues *Ile10|40|70* and *Leu16|46|76* for the stabilization of T3-785 is limited by the dispersive forces common to most residue-residue interactions: *Ile10-Ile40* (1.15 kcal/mol), *Leu16-Leu46* (-0.34 kcal/mol), *Leu16-Ala47 (-0.49* kcal/mol), *Leu16-Gly48 (-0.33* kcal/mol), *Leu46-Leu76* (-2.40 kcal/mol), *Leu46-Ala77* (0,15 kcal/mol), *Leu46-Gly78* (-0.26 kcal/mol), *Ile70-Ile10* (-0.62 kcal/mol), *Ile70-Ala13* (-1.58 kcal/mol), *Leu76-Ala17* (-1.76 kcal/mol), *Leu76-Pro22* (0.82 kcal/mol) - Figure 15.

More favorable interactions occur when hydrophobic contacts are replaced by induced charge-dipole forces promoted by *Arg14|44|74* (details will be presented below), water-mediated hydrogen bonds present in *Gly9-Ile40* (-10.92 kcal/mol), *Thr11-Ile40* (-31.39 kcal/mol), *Gly15-Leu46* (-9.66 kcal/mol), *Gly39-Ile70* (8.24 kcal/mol), *Thr41-Ile70* (-43.86 kcal/mol), *Gly45-Leu76* (-8.18 kcal/mol) and *Gly66-Ile10* (-12.17 kcal/mol) and Hyp68-Ile10 (-47.50 kcal/mol) or even unconventional hydrogen bonds observed in *Gly39-Ile10* (-5,96 kcal/mol), *Gly45-Leu16* (-3.88 kcal/mol), *Gly69-Ile40* (-3.83 kcal/mol), *Gly75- Leu46* (-4.20 kcal/mol), *Gly12-Ile70* (-3.29 kcal/mol) and *Gly18-Leu76* (-7.98 kcal/mol) - Figure 19, 20 and 21.

The hydrogen bonds observed here are consistent with those described in crystals with the *Leu - Hyp - Gly* triad (OKUYAMA et al., 2007); and also in those systems with the *Gly-Leu-Leu* sequence, which in addition to having slightly stabilizing *Van der Walls* Interactions, have a solvent-accessible region that is susceptible to interactions (PERSIKOV et al., 2002).

Figura 19 - Interchain bond energies of the aliphatic residues (a) Ile and (b) Leu within a radius of 8^a and considering the direction A->B, B->C and C->A.

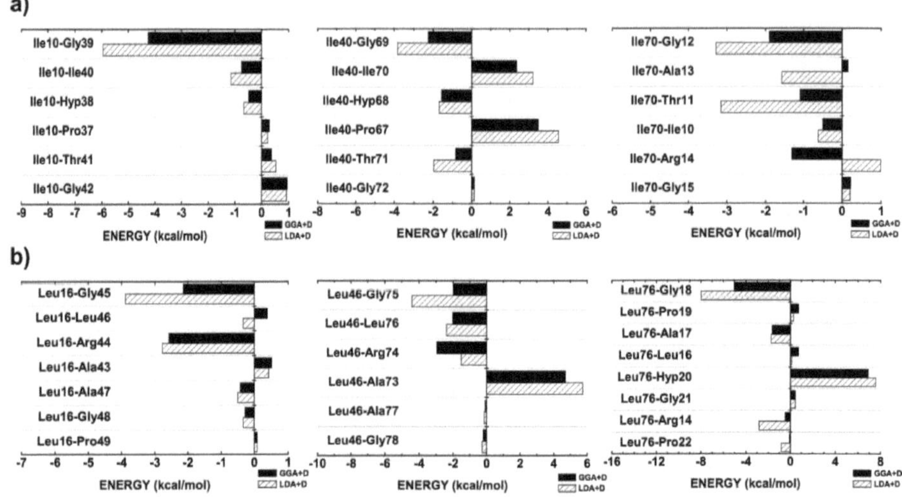

Source: *Prepared by the author.*

Figura 20 - Pattern of water-mediated hydrogen bonds involved in the most effective interchain interactions of Leu/Ile residues.

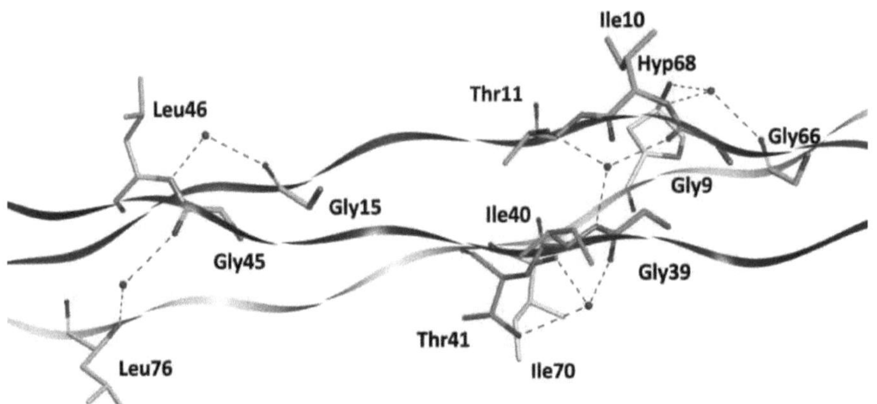

Source: *Prepared by the author.*

Figura 21 - Pattern of unconventional hydrogen bonds involved in important interchain interactions of Leu/Ile residues.

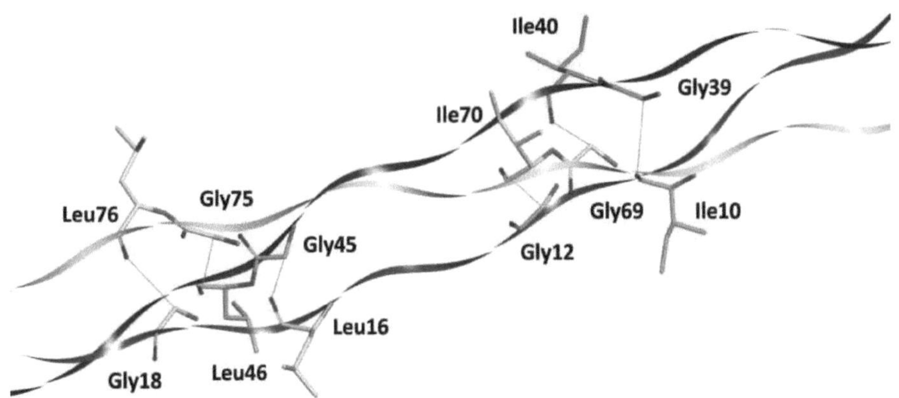

Source: *Prepared by the author.*

Glycine

The different types of collagen, even with their structural and functional peculiarities, are formed by repeats of the *Gly-Xaa-Yaa* sequence (triad). The *Gly* residues are essential for the formation of the triple helix, as they alone are small enough to occupy the core (interior) of the helix without a strong steric hindrance (BHOWMICK; FIELDS, 2013). Even though it doesn't have a charged (polar) side chain like the amino acid *Arg* (*Hyp* and *Thr*) capable of interacting through ion-dipole bonds, *Gly* has relatively strong intermolecular interactions. The (average) energies of *Gly-AB*, *Gly-BC* and *Gly-CA* are -18.99 kcal/mol, -17.02 kcal/mol and -18.81 kcal/mol. When we consider these values in the context of the triple helix as a whole (*ABC* system), we find that the *Gly* residues are responsible for 7.80% of the average interaction between the helices, which makes the glycine amino acid the second most important individual aliphyl residue. As we shall see, the CO carbonyl groups (NH and C^a H groups) act as acceptor (donor) elements in well-defined patterns of hydrogen bonds, which are essential for the consistency of the collagen triple helix.

2.5.4 High and low stability points

Collagen is an oligopeptide made up of three α-chains intertwined together around a common axis, giving it a helical conformation with high structural consistency. As we've seen so far, the different types of amino acids play a part in the solidity of the system to varying degrees, depending on the intermolecular forces they exert. After equating all the residue-residue contacts, the 15 residues with the strongest (weakest) interactions were identified. Subsequently, these were scaled in the three-dimensional structure of the T3-785 peptide, in order to visualize the points of high (low) stability - Figure 22.

This figure shows the importance of intermolecular interactions in maintaining the conformational stability of collagen. Only two (Ala17 and Ala47) of the 15 main residues do not have a polar side chain, but they do have dipole-dipole interactions (_Ala47_ :C^a H-OC: _Leu76_). On the other hand, all the residues with less attractive interactions are essentially non-polar, especially the C-/N-terminal proline residues, which make up 50% of them.

Figure 22 - The 15 elements with the highest (red) and lowest (blue) interchain interaction energies.

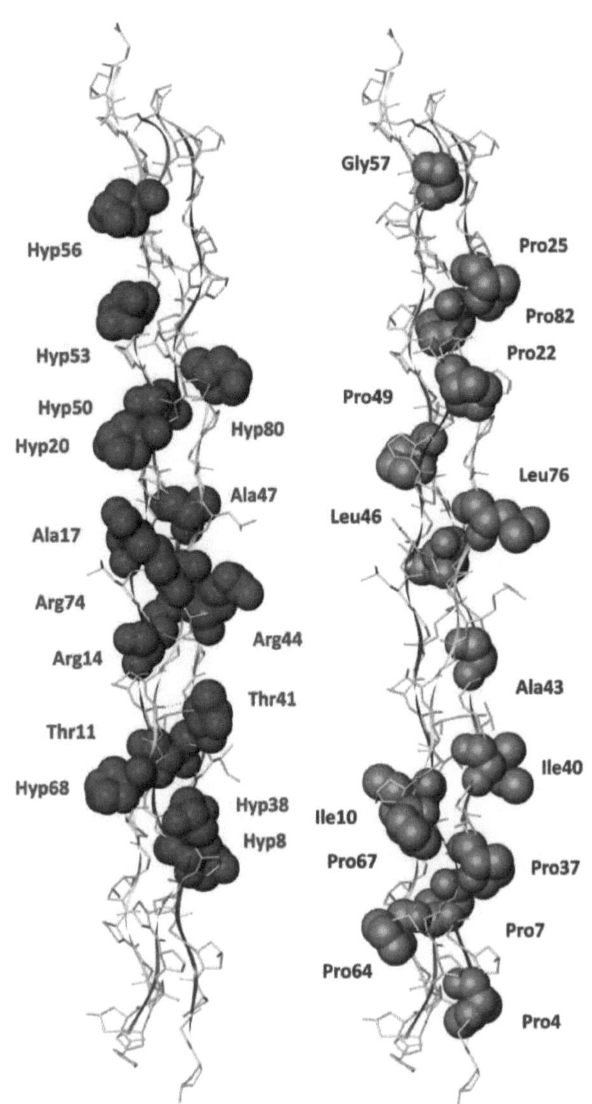

2.5.5 Hydrogen bonding patterns

In T3-785, the *Rich and Crick II* (RICH; CRICK, 1961) hydrogen bond pattern, <u>*Gly*:NH-OC:*Xaa*</u>, is present in all the *Gly-Xaa- Yaa* triads. Specifically, 25 of the 156 *Gly-residue* type interactions calculated here (excluding the terminal glycines) showed this pattern, precisely those that were most attractive. In this case: *Gly6-*

Pro34 (-4.08 kcal/mol), *Gly9-Pro37* (-6.54 kcal/mol), *Gly12-Ile40* (-5.63 kcal/mol), *Gly15-Ala43* (-1.26 kcal/mol), *Gly18-Leu46* (-7.63 kcal/mol), *Gly21-Pro49* (-7.79 kcal/mol), *Gly24- Pro52* (-11,65 kcal/mol), *Gly27-Pro55* (-8.94 kcal/mol) in the *AB* subsystem, *Gly36-Pro64* (-5.80 kcal/mol), *Gly39-Pro67* (-8.88 kcal/mol), *Gly42-Ile70* (10.70 kcal/mol), *Gly45-Ala73* (-4.73 kcal/mol), *Gly48-Leu76* (-9,58 kcal/mol), *Gly51-Pro79* (-4.16 kcal/mol), *Gly54-Pro82* (-6.23 kcal/mol), *Gly57-Pro85* (5.10 kcal/mol) in the *BC* subsystem, and *Gly66-Pro7* (-5.20 kcal/mol), *Gly69-Ile10* (-6.80 kcal/mol), *Gly72-Ala13* (-10.80 kcal/mol), *Gly75-Leu16* (-6.57 kcal/mol), *Gly78-Pro19* (-3.96 kcal/mol), *Gly81-Pro22* (-5.73 kcal/mol), *Gly84-Pro25* (4.87 kcal/mol), *Gly87-Pro28* (-4.16 kcal/mol) in the *CA* subsystem. In this peptide, the average distance (angle) from *Gly*:N to *Xaa*:CO is approximately 2.88 Å (166°) (KRAMER et al., 2001). Their importance is evident when we observe that the Gly-residue interactions with and without hydrogen bonds have average values of -6.57 kcal/mol and -2.15 kcal/mol, respectively.

A second network of hydrogen bonds is observed in the imino acid-free region (central zone) of the T3-785 peptide. In it, water molecules connect the carbonyl groups of the glycine residues of a chain with the amide groups (N-H) of residues located in the *Xaa* position and in adjacent chains - Figure 23. These interactions mediated by water molecules, <u>*Xaa*:NH-HOH-OC:Gly</u>, are observed in the contacts *Gly9-Ile40* (-10.92 kcal/mol), *Gly12-Ala43* (-16.01 kcal/mol) and *Gly15-Leu46* (9.66 kcal/mol) in the *AB* subsystem, *Gly39-Ile70* (-8.24 kcal/mol) and *Gly45-Leu76* (-8,18 kcal/mol) in the *BC subsystem*, as well as *Gly66-Ile10* (-12.17 kcal/mol), *Gly69-Ala13* (-7.56 kcal/mol) and *Gly72-Leu16* (-7.95 kcal/mol) in the *CA* subsystem - Figure 24 and 25. Unlike the observations of (KRAMER et al., 2001) regarding the crystallographic data of T3-785 (1BKV), the *Ala73*:NH-HOH-OC:*Gly42* hydrogen bond does not exist, even because the water molecule closest to the NH group of *Ala73* is at a distance of 6.43 Å. The absence of this contact justifies the low energy of the *Gly42 - Ala73* bond (-1.01 kcal/mol). The results confirm the hypothesis that this second network of hydrogen bonds strengthens the triple-helix conformation in regions of collagen poor in imino acids.

In three situations, water molecules not only enable *Xaa* :NH- HOH-OC: *Gly, but* also <u>*Ile* :NH-HOH-OH: *Yaa*</u> type contacts. The molecules responsible for *Ile40* :NH-

HOH-O: *Gly9* and *Ile70* :NH-HOH- OC:*Gly39* make an additional contact with the OH (hydroxyl) group of *Thr11* and *Thr41*. The importance of *Ile40*:NH-HOH-OH:*Thr11* (*Ile70*:NH-HOH-OH:*Thr41*) is expressed in the value of -31.29 kcal/mol (-43.86 kcal/mol) for the binding energy of *Thr11* and *Ile40* (*Thr41* and *Ile70*), in this case the third (first) most attractive in the *AB* (*BC*) subsystem. In a final situation, the water between *Gly66* and *Ile10* also interacts with *Hyp68* (*Ile10*:NH-HOH-OH:Hyp68*), directly influencing the *Hyp68-Ile10* attraction (-47.50, second most attractive in *CA*).

In addition to this web of bonds, the unconventional hydrogen bonds, <u>*Gly* :C$^{\alpha}$ H-OC: *Xaa(Ile/Leu/Pro/Hyp)*</u>, in the integrity of collagen were evaluated in terms of energy - Figure 24. In view of this, we can see that there are <u>six </u>contacts involving the <u>*Ile/Leu and Gly* </u>residues - *Gly39*:C$^{\alpha}$ *H-OC:Ile10* (-5.96 kcal/mol), *Gly45*:C$^{\alpha}$ H-OC:*Leu16* (3.88 kcal/mol), *Gly69*:C$^{\alpha}$ *H-OC:Ile40* (-3.83 kcal/mol), *Gly75*:C$^{\alpha}$ H-OC:*Leu46* (4.42 kcal/mol), *Gly12*:C$^{\alpha}$ *H-OC:Ile70* (-3.29 kcal/mol) and *Gly18*:C$^{\alpha}$ H-OC:*Leu76* (-7.98 kcal/mol). In addition to these, <u>15 *Pro* </u>interactions <u>with *Gly* </u>have this differentiated hydrogen bond - Pro4-Gly33 (-5.09 kcal/mol), Pro7- Gly36 (-6.12 kcal/mol), Pro19-Gly48 (-8.03 kcal/mol), Pro22-Gly51 (-4.79 kcal/mol), Pro25-Gly54 (-6.58 kcal/mol), Pro34-Gly63 (-4.81 kcal/mol), Pro37-Gly66 (-4.48 kcal/mol), Pro49-Gly78 (-6.13 kcal/mol), Pro52-Gly81 (-5.41 kcal/mol), Pro55-Gly84 (-7.00 kcal/mol), Pro64-Gly6 (-6,61 kcal/mol), Pro67-Gly9 (-3.00 kcal/mol), Pro79-Gly21 (-4.72 kcal/mol), Pro82-Gly24 (-7.39 kcal/mol) and Pro85-Gly27 (-3.91 kcal/mol).

In addition to these, we highlight the <u>14 ProO H-O$^{\gamma\delta}$:*Hyp* </u>involving glycine and proline residues, with other proline and hydroxyproline from adjacent chains, respectively - Hyp5-Pro34 (-20.66 kcal/mol), Hyp8- Pro37 (-27.54 kcal/mol), Hyp20-Pro49 (-41.01 kcal/mol), Hyp23-Pro52 (11.77 kcal/mol), Hyp26-Pro55 (-14.34 kcal/mol), Hyp35-Pro64 (-16.54 kcal/mol), Hyp38-Pro67 (-22.42 kcal/mol), Hyp50-Pro79 (-20.62 kcal/mol), Hyp53-Pro82 (-19.19 kcal/mol), Hyp56-Pro85 (-25,15 kcal/mol), Hyp65-Pro7 (-17.52 kcal/mol), Hyp80-Pro22 (-26.48 kcal/mol), Hyp83-Pro25 (-13.86 kcal/mol) and Hyp86-Pro28 (-10.81 kcal/mol).

Finally, <u>two *Ala*:C$^{\alpha}$ H-OC:*Leu* </u>bonds were observed, one in Ala17-Leu46 (-23.05 kcal/mol) and the other in Ala47-Leu76 (-13.76 kcal/mol). The existence and importance of this type of interaction for the collagen triple helix was first discussed by Bella and Berman (1996). The network of interchain C$^\alpha$ H-OC interactions acts to align the three chains and thus cooperatively lowers the total energy of the system.

Still on the subject of these data, we note that hydrogen bonds - whether conventional, water-mediated or of the unconventional type - are present in 62 residue-residue interactions, together this amount is responsible for a joint interchain bond energy of -768.94 kcal/mol (164.25+80.69+122.65+401.35), which corresponds to approximately 44% of the total interaction energy of the triple helix.

Figure 23 - Appearance of water-mediated hydrogen bonds in the crystallographic structure of T3-785.

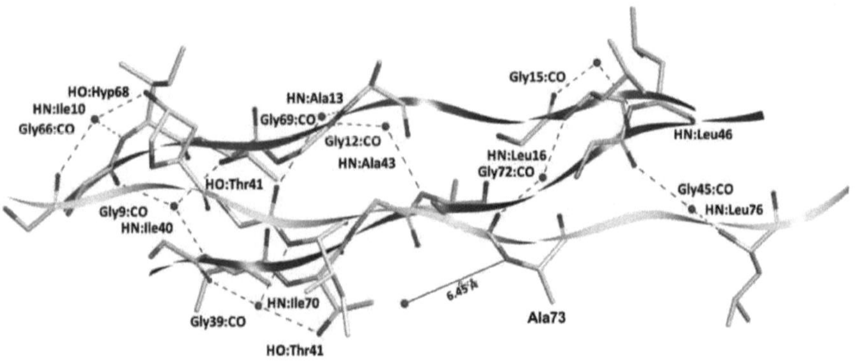

Source: *Prepared by the author.*

Figure 24 - 2D view of the hydrogen bonding patterns of T3-785. (a) Conventional hydrogen bonds and those mediated by water - Gly:NH- OC:Ile/Leu/Thr, Xaa:NH-HOH-OC:Gly + Ile:NH-HOH-OH:Yaa - are represented by the blue and red dashed lines, respectively. (b) Gly:C^a H-OC:Xaa are said to be unconventional and are shown in orange.

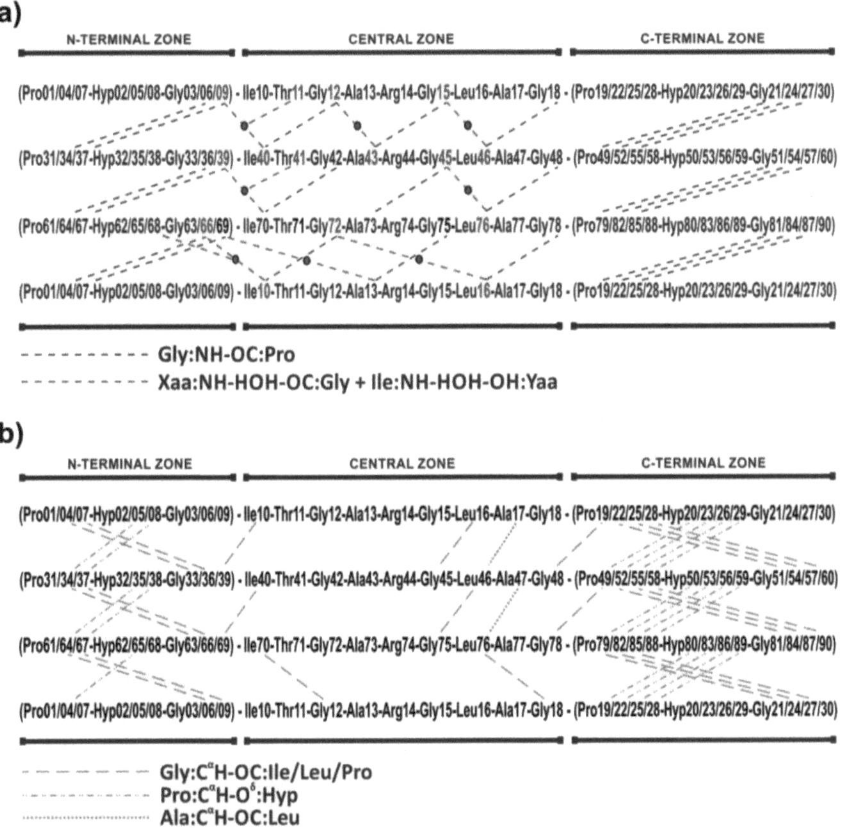

Source: *Prepared by the author.*

2.6 Collagen - a new perception

In order to clarify the structural and energetic aspects of the stability of the human collagen triple helix, the interaction energies between the amino acid residues that make up the homotrimeric peptide T3-785 were quantified using quantum mechanics calculations in the light of Density Functional Theory (DFT) and the Molecular Fractionation with Conjugated Hoods (MFCC) method. Thus, the forces of attraction and repulsion of each of the 90 amino acid residues that make up the system were signed and the regions with high/low stability identified and comparatively analyzed.

From the data analyzed here, the zones that make up T3-785 have the following organization, in terms of stabilizing importance: N-terminal < C-terminal < Central. The *Gly-Xaa-Yaa* structural blocks support the conformational integrity of the system in the order *Leu-Ala-Gly < Ile-Thr-Gly < Pro-Hyp-Gly < Ala-Arg-Gly*.

This work also assessed the individual relevance/importance of each type of T3-785 residue for stabilizing the system. From this perspective, the quality of each amino acid residue was not determined by the number of times it appeared in the structure, but by the strength with which it acted individually in the system. Therefore, the order of relevance was observed: *Arg* (39.5% relative individual influence) > *Hyp* (17.60%) > *Thr* (16.43%) > *Ala* (11.13%) > *Gly* (7.80%) > *Pro* (3.50%) > *Leu* (2.17%) > *Ile* (1.87%).

The interchain interaction energies of the amino acid residues showed varying degrees of attractiveness depending on the chemical nature of their side chain and the microsolvation environment surrounding them, factors which are crucial for the non-covalent interactions they make. The high attractiveness of the *Arg14*, *Arg44* and *Arg74* residues indicates the importance of permanent/induced charge-dipole contacts for the stability of the triple helix. In addition to these, conventional hydrogen bonds (*Gly*:NH-OC:*Xaa*) also stand out, as they are present in the strongest interactions involving the Gly residues.

The so-called non-conventional hydrogen bonds (*Gly* :C^a H-OC: *Xaa*) together with those mediated by water molecules (*Xaa* :NH-HOH- OC:*Gly*) are mainly responsible for the attraction of many interchain residues, especially those with the participation of the carbonyl/amine/hydroxyl groups of Gly/Ile/Thr residues, respectively. The energies of the residue-residue systems that showed water-mediated hydrogen interactions suggest that solvent effects play an important indirect role in the stabilization of T3-785. The dominant indirect role of the solvent in directing non-covalent bonds was previously discussed by Baron et al. (2010).

There is a strong difference between the interaction energies of the *Pro* and *Hyp* residues evaluated in this study. The increase in polarity promoted by the addition of the hydroxyl favors stronger stabilizing non-covalent contacts than those observed in the residue-residue interactions of the prolines. This perspective does not refute the idea that hydroxyproline residues favor the formation and preservation of the collagen triple helix through stereoelectronic effects (KOTCH; GUZEI; RAINES, 2008; BETSCHER et al., 2001; VITAGLIANO et al., 2001).

It was found that aliphatic residues also have reasonably attractive interactions - the *Ala*, *Leu*, *Ile* and *Gly* residues present in T3-785 together account for 22.97% of the average attractiveness of the triple helix. We therefore corroborate the hypothesis that the process of self-organization in triple helixes is also driven by *Van der Walls* interactions considered to be "weak" (OKUYAMA et al., 2007).

The low average attractiveness and equivalence between the *Ile* and *Leu* residues is due to the chemical nature of their respective radicals. In addition to being structurally similar, the aliphatic butyl functional groups stabilize the collagen triple helix through essentially hydrophobic contacts. As stronger electrostatic interactions occur in pairs containing these amino acids, the attractions become more favorable.

The alanine residues that occupy the Yaa position of the Gly-Xaa- Yaa triad of collagen are individually more relevant than those located in Xaa, since the latter are capable of making strong hydrophobic contacts with proline and leucine residues.

From all the above, it can be seen that quantum computational calculations were adequate in the quantitative evaluation of the relative contributions of each amino acid residue to the stabilization of collagen and, finally, reinforced the importance of non-covalent polar contacts.

The effectiveness of the methodology used (for the first time) in this book has led to the emergence of new applications for molecular fractionation techniques. Currently, the energetic link related to the complexation of synthetic collagen (Gly-Pro-HypP-Gly-Phe-Hyp-Gly-Glu- Arg-(Gly-Pro-Hyp)3 with the α2β1 domain of integrin has been evaluated using the MFCC methodology. New projects are being developed to analyze the stability of peptides with different secondary structure patterns and structural motifs - helix-π, helix-310, beta-sheet, zinc finger, α-turn-α, β-α-β.

In addition, the pioneering study of the conformational stability in energy terms of an amino acid-free region will prompt research into the development of artificial, highly stable collagens through bioengineering. The prospect is that the interactional data will support the choice of amino acids that will make up poly(Gly-Xaa- Yaa)$_{n\text{-type}}$ polymers.

3 FINAL CONSIDERATIONS

In the book entitled *Deciphering the Structural Stability of Human Collagen,* the crystal structure of the homotrimeric peptide T3-785 (PDB: 1BKV) was analyzed in terms of the energetic aspects of affinity between its monomers. In this case, the interaction energies between the chains of the triple helix were calculated in order to formulate predictions related to its structural integrity. In this way, we advanced the discussion on the stability of the collagen triple helix based on a structurally related model peptide.

4 REFERENCES

ABOU NEEL, E. A. et al. Collagen-emerging collagen-based therapies hit the patient. **Advanced drug delivery reviews,** v. 65, n. 4, p. 429-456, 2013.

ALBERTS, B. et al. Figure 19-47: The intracellular and extracellular events in the formation of a collagen fibril. In: Molecular Biology of the Cell. 4. ed. New York: Garland Science, 2002.

ALBUQUERQUE, E. L. et al. DNA-based nanobiostructured devices: The role of quasiperiodicity and correlation effects. **Physics Reports**, v. 535, p. 139-209, 2014.

AL-MOUSILLY, M. M.; ALAJELI, I. S.; ABDULRAHMAN, L. K. Study the healing effect of collagen hydrolysate for the treatment of bone tail fracture in mice. **International Journal of Pharmacy and Pharmaceutical Sciences**, v. 6, n. 6, p. 67-71, 2014.

ANDERSON, D. W. et al. A glycine (415)-to-serine substitution results in impaired secretion and decreased thermal stability of type III procollagen in a patient with Ehlers- Danlos syndrome type IV. **Human Mutation**, v. 9, n. 1, p. 62-63, 1997.

ANDRICOPULO, A. D.; MONTANARI, C. A. Structure-Activity relationships for the design of small-molecule inhibitors. **Mini-Reviews in Medicinal Chemistry**, v. 5, n. 6, p. 585-593, 2005.

ANTONY, J. et al. Protein-ligand interaction energies with dispersion corrected density functional theory and high-level wave function based methods. **Journal of Physical Chemistry A**, v. 115, n. 41, p. 11210-11220, 2011.

ANTONY, J.; GRIMME, S. Fully ab initio protein-ligand interaction energies with dispersion corrected density functional theory. **Journal of Computational Chemistry**, v. 33, n. 21, p. 1730-1739, 2012.

ASZÓDI, A. et al. What mouse mutants teach us about extracellular matrix function. **Annual Review of Cell and Developmental Biology**, v. 22, p. 591-621, 2006.

BAGCHI, D. et al. Effects of orally administered undenatured type II collagen against arthritic inflammatory disease: a mechanistic exploration. **International Journal of Clinical Pharmacology Research**, v. 22, n. 3-4, p. 101-110, 2002.

BAILEY, N. P.; SETHNA, J. P. Macroscopic measure of the cohesive length scale: Fracture of notched single-crystal silicon. **Physical Review B**, v. 68, n.20, p. 205204-1-8, 2003.

BANN, J. G.; BACHINGER, H. P. Glycosylation/hydroxylation-induced stabilization of the collagen triple helix. 4-trans-hydroxyproline in the Xaa position can stabilize the triple helix. **Journal of Biological Chemistry**, v. 275, n. 32, p. 24466-24469, 2000.

BARON, R. et al. Water in Cavity-Ligand Recognition. **Journal of the American Chemical Society,** v. 132, n. 34, p. 12091-12097, 2010.

BARREIRO, E. J.; FRAGA, C. A. M. **Medicinal Chemistry: The Molecular Basis of Drug Action**. Artmed Editora: Porto Alegre, 2001.

BASIUK, V. A.; HENAO-HOLGUiN, L. V. Dispersion-corrected density functional theory calculations of meso-tetraphenylporphine-c60 complex by using DMol3 module. **Journal of Computational and Theoretical Nanoscience**, v. 11, n. 7, p. 1609-1615, 2014.

BELLA, J. et al. Crystal and molecular structure of a collagen-like peptide at 1.9 Å resolution. **Science**, v. 266, n. 5182, p. 75-81, 1994.

BELLA, J.; BRODSKY, B.; BERMAN, H. M. Hydration structure of a collagen peptide. **Structure**, v. 3, n. 9, p. 893-906, 1995.

BELLA, J.; BERMAN, H. M. Crystallographic evidence for C alpha-H...O=C hydrogen bonds in a collagen triple helix. **Journal of Molecular Biology**, v. 264, n. 4, p. 734-742, 1996.

BELLA, J. A new method for describing the helical conformation of collagen: dependence of the triple helical twist on amino acid sequence. **Journal of Structural Biology**, v. 170, n. 2, p. 377-391, 2010.

BERG, R. A.; PROCKOP, D. J. The thermal transition of a non-hydroxylated form of collagen. Evidence for a role for hydroxyproline in stabilizing the triple-helix of collagen. **Biochemical and Biophysical research Communications**, v. 52, n. 1, p. 115-120, 1973.

BERISIO, R. et al. Crystal structure of a collagen-like polypeptide with repeating sequence pro-hyp-gly at 1.4 Å resolution: Implications for collagen hydration. **Biopolymers**, v. 56, n. 1, p. 8-13, 2000.

BERISIO, R. et al. Crystal structure of the collagen triple helix model [(Pro-Pro-Gly)10]3. **Protein Science**, v. 11, n. 2, p. 262-270, 2002.

BERISIO, R. et al. Imino acids and collagen triple helix stability: Characterization of collagen-like polypeptides containing Hyp-Hyp-Gly sequence repeats. **Journal of the American Chemical Society**, v. 126, n. 37, p. 11402-11403, 2004.

BERISIO, R. et al. Role of side chains in collagen triple helix stabilization and partner recognition. **Journal of Peptide Science**, v. 15, n. 3, p. 131-140, 2009.

BEZERRIL, L. M. et al. Current voltage characteristics of double-strand DNA sequences. **Physics Letters A** (Print), v. 373, p. 3381-3385, 2009.

BEZERRIL, L. M. et al. Charge transport in fibrous/not fibrous α3-helical and (5Q,7Q)α3 variant peptides. **Applied Physics Letters**, v. 98, p. 053702, 2011.

BETSCHER, L. E. et al. Conformational stability of collagen relies on a stereoelectronic effect. **Journal of the American Chemical Society**, v. 123, n. 4, p. 777-778, 2001.

BHATNAGAR, R.S. et al. Interchain proline:proline contacts contribute to the stability of triple-helical conformation. **Journal of Biomolecular Structure & Dynamics**, v. 6, n. 2, p. 223-233, 1988.

BHOWMICK, M.; FIELDS, G. B. Stabilization of Collagen-Model, Triple-Helical Peptides for In Vitro and In Vivo Applications. **Methods in Molecular Biology**, v. 1081, p. 167-194, 2013.

BISSANTZ, C.; KUHN, B.; STAHL, M. A medicinal chemist's guide to molecular interactions. **Journal of Medicinal Chemistry**, v. 53, n. 14, p. 5061-5084, 2010. Review. Erratum in: Journal of Medicinal Chemistry,v. 53, n. 16, 6241, 2010.

BONADIO, J.; BYERS, P. H. Subtle structural alterations in the chains of type I procollagen produce osteogenesis imperfecta type II. **Nature**, v. 316, n. 6026, p. 363-366, 1985.

BRODSKY, B.; PERSIKOV, A. V. Molecular structure of the collagen triple helix. **Advances in Protein Chemistry**, v. 70, 301-339, 2005.

BRUYÈRE, O. et al. Effect of collagen hydrolysate in articular pain: a 6-month randomized, double-blind, placebo controlled study. **Complement Therapies in Medicine**, v. 20, n. 3, p. 124-130, 2012.

BUEHLER, M. J. Atomistic and continuum modeling of mechanical properties of collagen: Elasticity, fracture, and self-assembly. **Journal of Materials Research**, v. 21, n. 8, p. 1947-1961, 2006.

BUEHLER, M. J. Nanomechanics of collagen fibrils under varying cross-link densities: Atomistic and continuum studies. **Journal of the Mechanical behavior of Biomedical Materials**, v. 1, n. 1, p. 59-67, 2008.

BULTINCK, P.; TOLLENAERE, J. P.; WINTER, H. **Computational medicinal chemistry for drug discovery**. New York: Marcel Dekker, p. 794, 2003.

BYERS, P. H.; COLE, W. G. Osteogenesis Imperfecta. In: ROYCE, P. M.; STEINMANN, B. (Org.), **Connective tissue and its hereditable disorders**. New York: Wiley-Liss, 2002, p. 385-430.

CAPRASECCA, S. et al. Geometry optimization in polarizable QM/MM models: the induced dipole formulation. **Journal of Chemical Theory and Computation**, v. 10, n. 4, p. 1588-1598, 2014.

CHARPENTIER, J. C. The triplet "molecular processes-product-process" engineering: The future of chemical engineering? **Chemical Engineering Science**, v. 57, n. 22-23, p. 46674690, 2002.

CEN, L. et al. Collagen tissue engineering: development of novel biomaterials and applications. **Pediatric research** v.63, n. 5, p. 429-496, 2008.

CHAKRABARTI, P.; BHATTACHARYYA, R. Geometry of nonbonded interactions involving planar groups in proteins. **Progress in Biophysics and Molecular Biology**, v. 95, n. 1-3, p. 83-137, 2007.

CHEN, X. H.; ZHANG, J. Z. Theoretical method for full ab initio calculation of DNA/RNA- ligand interaction energy. **Journal of Chemical Physics**, v. 120, n. 24, p. 11386-11391, 2004.

CHENG, H. et al. Location of Glycine Mutations within a Bacterial Collagen Protein Affects Degree of Disruption of Triple-helix Folding and Conformation. **Journal of Biological Chemistry**, v. 286, n. 3, p. 2041-2046, 2011a.

CHENG, W. et al. The content and ratio of type I and III collagen in skin differ with age and injury. **African Journal of Biotechnology**, v. 10, n. 13, p. 2524-2529, 2011b.

CLARK, K. L. et al. 24-Week study on the use of collagen hydrolysate as a dietary supplement in athletes with activity-related joint pain. **Current Medical Research and Opinion**, v. 24, n. 5, p. 1485-1496, 2008.

CLOSE, D. M. Calculated vertical ionization energies of the common α-amino acids in the gas phase and in solution. **The Journal of Physical Chemistry A**, v. 115, n. 13, p. 29002912, 2011.

COLE, W. G. Collagen genes: mutations affecting collagen structure and expression. **Progress in Nucleic Acid Research Molecular Biology**, v. 47, p. 29-80, 1994.

da COSTA, R. F. et al. Explaining statin inhibition effectiveness of HMG-CoA reductase by quantum biochemistry computations. **Physical Chemistry Chemical Physics**, v. 14, p. 1389-1398, 2012.

DAI, N.; WANG, X. J.; ETZKORN, F. A. The Effect of a Trans-Locked Gly- Pro Alkene Isostere on Collagen Triple Helix Stability. **Journal of the American Chemical Society**, v. 130, n. 16, p. 5396-5397, 2008.

DALGLEISH, R. The human collagen mutation database 1998. **Nucleic Acids Research**, v. 26, n. 1, p. 253-255, 1998.

DELLEY, B. An All-Electron Numerical Method for Solving the Local Density Functional for Polyatomic Molecules. Journal of Chemical Physics, v. 92, n. 1, p. 508-517, 1990.

DEREWENDA, Z. S.; LEE, L.; DEREWENDA, U. The occurrence of C-HO hydrogen bonds in proteins. **Journal of Molecular Biology**, v. 252, p. 248-262, 1995.

Di LULLO, G.A. et al. Mapping the ligand-binding sites and disease-associated mutations on the most abundant protein in the human, type I collagen. **Journal of Biological Chemistry**, v. 277, p. 4223-4231,2002.

DOI, M. et al. Collagen-like triple helix formation of synthetic (Pro-Pro-Gly)10 analogues: (4(S)-hydroxyprolyl-4(R)-hydroxyprolyl-Gly)10, (4(R)-hydroxyprolyl-4(R)-hydroxyprolyl- Gly)10 and (4(S)-fluoroprolyl-4(R)-fluoroprolyl-Gly)10. **Journal of Peptide Science**, v. 11, n. 10, p. 609-616, 2005.

EMSLEY, J. et al. Structural basis of collagen recognition by integrin α2β1. **Cell**, v. 101, n. 1, p. 47-56, 2000.

EXPOSITO, J.; LETHIAS, C. Invertebrate and Vertebrate Collagens. In: KEELEY, F. W.;

MECHAM, R. P. (Org), **Evolution of Extracellular Matrix,** Berlin-Heidelberg: Springer, 2013, p. 39-72.

FALLAS, J. A.; GAUBA, V.; HARTGERINK, J. D. Solution structure of an ABC collagen heterotrimer reveals a single-register helix stabilized by electrostatic interactions. **Journal of Biological Chemistry**, v. 284, n. 39, p. 26851-26859, 2009.

FAN, P. et al. Backbone dynamics of (Pro-Hyp-Gly)10 and designed collagen-like triplehelical peptide by 15N NMR relaxation and hydrogen-exchange measurements. **Biochemistry**, v. 32, n. 48, p. 13299-13309, 1993.

FIELDS, G. B.; PROCKOP, D. J. Perspectives on the synthesis and application of triplehelical, collagen-model peptides. **Biopolymers**, v. 40, n. 4, p. 345-357, 1996.

FIELDS, G. B. Synthesis and biological applications of collagen-model triple-helical peptides. **Organic & Biomolecular Chemistry**, v. 8, p. 1237-1258, 2010.

FOX, R. et al. Spontaneous carotid-cavernous fistula in Ehlers-Danlos syndrome. **Journal Neurology, Neurosurgery, Psychiatry,** v. 51, n. 7, p. 984-986, 1988.

GAUBA, V.; HARTGERINK, J. D. Self-assembled heterotrimeric collagen triple helices directed through electrostatic interactions. **Journal of the American Chemical Society**, v. 129, n. 28, p. 2683-2690, 2007a.

GAUBA, V.; HARTGERINK, J. D. Surprisingly high stability of collagen abc heterotrimer: Evaluation of side chain charge pairs. **Journal of the American Chemical Society**, v. 129, n. 48, p. 15034-15041, 2007b..

GAUBA, V.; HARTGERINK, J. D. Synthetic Collagen Heterotrimers: Structural Mimics of Wild-Type and Mutant Collagen Type I. **Journal of the American Chemical Society**, v. 130, n. 23, p. 7509-7515, 2008.

GAUTIERI, A. et al. Hierarchical structure and nanomechanics of collagen microfibrils from the atomistic scale up. **Nano Letters**, v. 11, n. 2, p. 757-766, 2011.

GILCHRIST, D. et al. Large kindred with Ehlers-Danlos syndrome type IV due to a point mutation (G571S) in the COL3A1 gene of type III procollagen: low risk of pregnancy complications and unexpected longevity in some affected relatives. **American Journal of Medical Genetics**, v. 82, n. 4, p. 305-311, 1999.

GLOWACKI, J.; MIZUNO, S. Collagen scaffolds for tissue engineering. **Biopolymers**, v. 89, n. 5, p. 338-44, 2008.

GORDON, M. K.; HAHN, R. A. Collagens. **Cell Tissue Research**, v. 339, n. 1, p. 247-257, 2010.

GORDON, M. S. et al. Fragmentation methods: a route to accurate calculations on large systems. **Chemical Reviews**, v. 112, n. 1, p. 632-672, 2012.

GRANT, M. E. From collagen chemistry towards cell therapy - a personal journey. **International Journal of Experimental Pathology**, v. 88, n. 4, p. 203-214, 2007.

GRIMME, S. Semiempirical GGA-type density functional constructed with a long-range dispersion correction. **Journal of Computational Chemistry**, v. 27, n. 15, p. 1787-1799, 2006.

GRIMME, S. Supramolecular binding thermodynamics by dispersion-corrected density functional theory. **Chemistry - A European Journal**, v. 18, n. 32, p. 9955-9964, 2012.

GROVER, V. et al. Collagen: A review. **Indian Journal of Oral Science**, v. 1, p. 15-24, 2010.

GUIDO, R. V. C.; ANDRICOPULO, A. D.; OLIVA, G. Drug planning, biotechnology and medicinal chemistry: applications in infectious diseases. **Estudos Avançados**, v. 24, n. 70, p. 81-98, 2010.

HABIB, M. A. et al. Experimental and Numerical investigation of La2NiO4membranes for oxygen separation: geometry optimization and model validation. **Journal of Energy Resources Technology**, 2015.

HODGES, J. A.; RAINES, R. T. Stereoelectronic and steric effects in the collagen triple helix: Toward a code for strand association **Journal of the American Chemical Society**, v. 127, n. 45, p. 15923-15932, 2005.

HONGO, C. et al. Average Crystal Structure of (Pro-Pro-Gly) 9 at 1.0A Resolution. **Polymer Journal**, v. 33, n. 10, p. 812-818, 2001.

HONGO, C. et al. Repetitive interactions observed in the crystal structure of a collagen- model peptide, [(Pro-Pro-Gly)9]3. **Journal of biochemistry**, v. 138, n. 2, p. 135-144, 2005.

HORN, A. H.; CLARK, T. Multipole electrostatic potential derived atomic charges in NDDO-methods with spd-basis sets. **Journal of Molecular Modeling**, v. 13, n. 2, p. 381392, 2007.

HU, L. et al. Do quantum mechanical energies calculated for small models of proteinactive sites converge? **Journal Physical Chemical A**, v. 113, n. 43, p. 11793-11800, 2009.

HULMES, D. J. S. The collagen superfamily - diverse structures and assemblies. **Essays in Biochemistry**, v. 27, n. 27, p. 49-67, 1992.

IMPROTA, R.; BERISIO, R.; VITAGLIANO, L. Contribution of dipole-dipole interactions to the stability of the collagen triple helix. **Protein Science**, v. 17, n. 5, p. 955-961, 2008.

INADA, Y.; ORITA, H. Efficiency of numerical basis sets for predicting the binding energies of hydrogen bonded complexes: evidence of small basis set superposition error compared to Gaussian basis sets. **Journal of Computational Chemistry**, v. 29, n. 2, p. 225-232, 2008.

ISAEV, A. N. C-H-O, O-H-C, and C-H-C interactions in complexes of carbocations and carboanions. **Russian Journal of Inorganic Chemistry**, v. 58, n. 7, p. 817-823, 2013.

JIRAVANICHANUN, N. et al. Unexpected puckering of hydroxyproline in the guest triplets, Hyp-

Pro-Gly and Pro-allo-Hyp-Gly sandwiched between Pro-Pro-Gly sequence. **ChemBioChem**, v. 6, n. 7, p. 1184-1187, 2005.

JIRAVANICHANUN, N.; NISHINO, N.; OKUYAMA, K. Conformation of allo-Hyp in the Y position in the host-guest peptide with the Pro-Pro-Gly sequence: Implication of the destabilization of (Pro-allo-Hyp-Gly)10. **Biopolymers**, v. 81, n. 3, p. 225-233, 2006.

JOHNSON, P. H. et al. A COL3A1 glycine 1006 to glutamic acid substitution in a patient with Ehlers-Danlos syndrome type IV detected by denaturing gradient gel electrophoresis. **Journal of Inherited Metabolic Disease**, v. 15, n. 3, p. 426-430, 1992.

JÓNSDÓTTIR, S. Ó.; JORGENSEN, F. S.; BRUNAK, S. Prediction methods and databases within chemoinformatics: emphasis on drugs and drug candidates. **Bioinformatics**, v. 21, n. 10, 2145-2160, 2005.

KAWAHARA, K. et al. Effect of hydration on the stability of the collagen-like triple-helical structure of [4(R)-hydroxyprolyl-4(R)-hydroxyprolylglycine]10. **Biochemistry**, v. 44, n. 48, p. 15812-15822, 2005.

KOIDE, T.; NISHIKAWA, Y.; TAKAHARA, Y. Synthesis of heterotrimeric collagen models containing Arg residues in Y-positions and analysis of their conformational stability. **Bioorganic & Medical Chemistry Letters**, v. 14, n. 1, p. 125-128, 2004.

KONTUSAARI, S. et al. Substitution of aspartate for glycine 1018 in the type III procollagen (COL3A1) gene causes type IV Ehlers-Danlos syndrome: the mutated allele is present in most blood leukocytes of the asymptomatic and mosaic mother. **American Journal of Human Genetics**, v. 51, n. 3, p. 497-507, 1992.

KOTCH, F. W.; GUZEI, A. I.; RAINES, R. T. Stabilization of the Collagen Triple Helix by O-Methylation of Hydroxyproline Residues. **Journal of the American Chemical Society**, v. 130, n. 10, p. 2952-2953, 2008.

KRAMER, R. Z. et al. X-Ray crystallographic determination of a collagen-like peptide with the repeating sequence (Pro-Pro-Gly). **Journal of Molecular Biology**, v. 280, n. 4, p. 623-638, 1998.

KRAMER, R. Z. et al. Sequence dependent conformational variations of collagen triplehelical structure. **Nature Structural Biology**, v. 6, p. 454-457, 1999.

KRAMER, R. Z. et al. Staggered molecular packing in crystals of a collagen like peptide with a single charged pair. **Journal of Molecular Biology**, v. 301, n. 5, p. 1191-1205, 2000.

KRAMER, R. Z. et al. The crystal and molecular structure of a collagen-like peptide with a biologically relevant sequence. **Journal of Molecular Biology**, v. 311, n. 1, p. 131-147, 2001.

KROES, H. Y.; PALS, G.; van ESSEN, A. J. Ehlers-Danlos syndrome type IV: unusual congenital anomalies in a mother and son with a COL3A1 mutation and a normal collagen III protein profile.

Clinical Genetics, v. 63, n. 3, p. 224-227, 2003. Note: Erratum: Clinical Genetics, v. 64, p. 375, 2003.

LAUER-FIELDS, J. et al. Triple-helical transition state analogues: A new class of selective matrix metalloproteinase inhibitors. **Journal of the American Chemical** Society, v. 129, n. 34, p. 10408-10417, 2007.

LANIG, H. et al. Molecular dynamics simulations of the tetracycline-repressor protein: the mechanism of induction. **Journal of Molecular Biology**, v. 359, n. 4, p.1125-1136, 2006.

LEACH, A. R. **Molecular Modeling: Principles and Applications.** 2. ed., Pretince Hall, 2001.

LEE, J. Y. et al. New use of a three dimensional pellet culture system for human intervertebral disc cells: initial characterization and potential use for tissue engineering. **Spine**, v. 26, n. 21, p. 2316-2322, 2001.

LEE, C. H.; SINGLA, A.; LEE, Y. Biomedical applications of collagen. **International Journal of Pharmaceutics**, v. 221, n. 1-2, p. 1-22, 2001.

MA, B.; NUSSINOV, R. From computational quantum chemistry to computational biology: experiments and computations are (full) partners. **Physical Biology**, v. 1, n. 4, p. 23-26, 2004.

MANIKANDAN, K.; RAMAKUMAR, S. The occurrence of C--H...O hydrogen bonds in alpha-helices and helix termini in globular proteins. **Proteins**, v. 56, n. 4, p. 768-781, 2004.

MARLOWE, A. E.; SINGH, A.; YINGLING, Y. G. The effect of point mutations on structure and mechanical properties of collagen-like fibril: A molecular dynamics study. **Materials Science and Engeneering: C**, v. 32, n. 8, p. 2583-2588, 2012.

MARTINS, A. C. V. et al. An ab initio explanation of the activation and antagonism strength of an AMPA-sensitive glutamate receptor. **RSC Advances**, v. 3, n. 35, p. 14988-14992, 2013.

MCGRORY, J. et al. Abnormal extracellular matrix in Ehlers-Danlos syndrome type IV due to the substitution of glycine 934 by glutamic acid in the triple helical domain of type III collagen. **Clinical Genetics**, v. 50, n. 2, p. 442-445, 1996.

MCGRORY, J.; COSTA, T.; COLE, W. G. A novel G499D substitution in the alpha-1(III) chain of type III collagen produces variable forms of Ehlers-Danlos syndrome type IV. **Human Mutation**, v. 7, n. 1, p. 59-60, 1996.

MEI, Y. et al. Quantum study of mutational effect in binding of efavirenz to HIV-1 RT. **Proteins**, v. 59, n. 3, p. 489-495, 2005.

MENDES, G. A. et al. Electronic specific heat of an 3-helical polypeptide and its biochemical variants. **Chemical Physics Letters** (Print), v. 542, p. 123-127, 2012.

MENIKARACHCHI, L. C.; GASCÓN, J. A. QM/MM approaches in medicinal chemistry research. **Current Topics in Medicinal Chemistry**, v. 10, n. 1, p. 46-54, 2010.

MIZUNO, K.; HAYASHI, T.; BACHINGER, H. P. Hydroxylation-induced stabilization of the collagen triple helix. Further characterization of peptides with 4(R)-hydroxyproline in the Xaa position. **The Journal of Biological Chemistry**, v. 278, n. 34, p. 32373-32379, 2003.

MIZUNO, K. et al. Vascular Ehlers-Danlos syndrome mutations in type III collagen differently stall the triple helical folding. **Journal of Biological Chemistry**, v. 288, n. 26, p. 19166-19176, 2013.

MORADI, M. et al. Conformations and free energy landscapes of polyproline peptides. **Proceedings of the National Academy of Sciences**, v. 106, n. 49, p. 20746-20751, 2009.

MYLLYHARJU, J.; KIVIRIKKO, K. I. Collagens, modifying enzymes and their mutations in humans, flies and worms. **Trends in Genetics**, v. 20, n. 1, p. 33-34, 2004.

NAGARAJAN, V.; KAMITORI, S.; OKUYAMA, K. Structure analysis of a collagen-model peptide with a (Pro-Hyp-Gly) sequence repeat. **Journal of Biochemistry**, v. 125, n. 2, p. 310-318, 1999.

NARCISI, P. et al. Single base mutation that substitutes glutamic acid for glycine 1021 in the COL3A1 gene and causes Ehlers-Danlos syndrome type IV. **American Journal of Medical Genetics**, v. 46, n. 3, p. 278-283, 1993.

NARCISI, P. et al. A family with Ehlers-Danlos syndrome type III/articular hypermobility syndrome has a glycine 637-to-serine substitution in type III collagen. **Human Molecular Genetics**, v. 3, n. 9, p. 1617-1620, 1994.

NISHI, Y. et al. Different Effects of 4-Hydroxyproline and 4-Fluoroproline on the Stability of Collagen Triple Helix. **Biochemistry**, v. 44, n. 16, p. 6034-6042, 2005.

NUYTINCK, L. et al. Detection and characterization of an overmodified type III collagen by analysis of non-cutaneous connective tissues in a patient with Ehlers-Danlos syndrome IV. **Journal of Medical Genetics**, v. 29, n. 6, p. 375-380, 1992.

O'BRIEN, F. J. Biomaterials & scaffolds for tissue engineering. **Materials Today**, v. 14, n. 3, p. 88-95, 2011.

O'HAGAN, D. Understanding organofluorine chemistry. An introduction to the C-F bond. **Chemical Society Reviews**, v. 37, n. 2, p. 308-319, 2008.

OKUYAMA, K. et al. Crystal and molecular structure of a collagen-like polypeptide (Pro- Pro-Gly)10. **Journal of Molecular Biology**, v. 152, n. 2, p. 427-443, 1981.

OKUYAMA, K. et al. Crystal structures of collagen model peptides with Pro-Hyp-Gly repeating sequence at 1.26 Å resolution: Implications for proline ring puckering. **Biopolymers**, v. 76, n. 5, p. 367-377, 2004.

OKUYAMA, K. et al. Unique side chain conformation of a Leu residue in a triple-helical structure. **Biopolymers**, v. 86, n. 3, p. 212-221, 2007.

OKUYAMA, K. et al. Stabilization of triple-helical structures of collagen peptides containing a Hyp-

Thr-Gly, Hyp-Val-Gly, or Hyp-Ser-Gly sequence. **Biopolymers**, v. 95, n. 9, p. 628640, 2011.

OKUYAMA, K. et al. Crystal structure of (Gly-Pro-Hyp) 9: Implications for the collagen molecular model. **Biopolymers**, v. 97, n. 8, p. 607-616, 2012.

OLIVEIRA, J. I. N., et al. Conductance of single microRNAs chains related to the autism spectrum disorder. **Europhysics Letters**, v. 107, n. 6, p. 68006, 2014a.

OLIVEIRA, J. I. N., et al. Electronic transport through oligopeptide chains: An artificial prototype of a molecular diode. **Chemical Physics Letters**, v. 612, p. 14-19, 2014b.

ORTMANN, F.; BECHSTEDT, F.; SCHMIDT, W. G. Semiempirical van der Waals correction to the density functional description of solids and molecular structures. **Physical Review B**, v. 73, p. 205101, 2006.

OTTL, J. et al. Recognition and catabolism of synthetic heterotrimeric collagen peptides by matrix metalloproteinases. **Chemistry & Biology**, v. 7, n. 2, p. 119-132, 2000.

PADIYAR, G. S.; SESHADRI, T. P. Trifurcated (four-center) hydrogen bond in solid state crystal structure of 5-amino-5-deoxyadenosine p-toluenesulfonate. **Nucleosides and Nucleotides**, v. 15, n. 4, p. 857-865, 1996.

PALMERI, S. et al. Neurological presentation of Ehlers-Danlos syndrome type IV in a family with parental mosaicism. **Clinical Genetics**, v. 63, n. 6, p. 510-515, 2003.

PARENTEAU-BAREIL, R.; GAUVIN, R.; BERTHOD, F. Collagen-Based Biomaterials for Tissue Engineering Applications. **Materials**, v. 3, n. 3, p. 1863-1887, 2010.

PATI, F. et al. Collagen scaffolds derived from fresh water fish origin and their biocompatibility. **Journal of Biomedical Materials Research Part A**, v. 100, n. 4, p. 1068-1079, 2012.

PEPIN, M. et al. Clinical and genetic features of Ehlers-Danlos syndrome type IV, the vascular type. **New England Journal of Medicine**, v. 342, p. 673-680, 2000. Note: Erratum: New England Journal of Medicine, v. 344, p. 392, 2001.

PERDEW, J. P.; BURKE, K.; ERNZERHOF, M. Generalized gradient approximation made simple. **Physical review letters**, v. 77, n. 18, p. 3865-3868, 1996.

PERRET, S. et al. Unhydroxylated triple helical collagen I produced in transgenic plants provides new clues on the role of hydroxyproline in collagen folding and fibril formation. .**Journal of Biological Chemistry**, v. 276, n. 47, p. 43693-43698, 2001.

PERSIKOV, A. V. et al. Amino Acid Propensities for the Collagen Triple-Helix. **Biochemistry**, v. 39, n. 48, p. 14960-14967, 2000.

PERSIKOV, A. V.; RAMSHAW, J. A.; BRODSKY, B. Collagen model peptides: Sequence dependence of triple-helix stability. **Biopolymers**, v. 55, n. 6, p. 436-450, 2000.

PERSIKOV, A. V. et al. Peptide investigations of pairwise interactions in the collagen triple-helix. **Journal of Molecular Biology**, v. 316, n. 2, p. 385-394, 2002.

PROCKOP, D. J.; KIVIRIKKO, K. I. Heritable diseases of collagen. **The New England Journal of Medicine**, v. 311, n. 6, p. 376-386, 1984.

RAHA, K. et al. The role of quantum mechanics in structure-based drug design. **Drug Discovery Today**, v. 12, n. 17-18, p. 725-731, 2007.

RAMACHANDRAN, G. N.; BANSAL, M.; BHATNAGAR, R. S. A hypothesis on the role of hydroxyproline in stabilizing collagen structure. **Biochimica et Biophysica Acta**, v.322, n. 1, p. 166-171, 1973.

RAMACHANDRAN, G. N.; CHANDRASEKHARAN, R. Interchain hydrogen bonds via bound water molecules in the collagen triple helix. **Biopolymers**, v. 6, n. 11, p. 1649-1658, 1968.

RAMACHANDRAN, G. N.; KARTHA, G. Structure of collagen. **Nature**, v. 174, p. 269-270, 1954.

RAMACHANDRAN, G. N.; KARTHA, G. Structure of Collagen. **Nature**, v. 176, p. 593-595, 1955.

RAMSHAW, J. A.; SHAH, N. K.; BRODSKY, B. Gly-X-Y Tripeptide Frequencies in Collagen: A Context for Host-Guest Triple-Helical Peptides. **Journal of Structural Biology**, v. 122, n. 1-2, p.86-91, 1998.

RIBEIRO, T. C. S. et al. The quantum biophysics of the isoniazid adduct NADH binding to its InhA reductase target. **New Journal of Chemistry**, v. 38, n. 7, p. 2946-2957, 2014.

RICARD-BLUM, S. The Collagen Family. **Cold Spring Harbor Perspectives in Biology**, v. 3, n. 1, p. a004978, 2011.

RICH, A.; CRICK, F. H. C. The molecular structure of collagen. **Journal of Molecular Biology**, v. 3, p. 483-506, 1961.

RICHARDS, A. J. et al. A 27-bp deletion from one allele of the type III collagen gene (COL3A1) in a large family with Ehlers-Danlos syndrome type IV. **Human Genetics**, v. 88, n. 3, p. 325-330, 1992.

RODRIGUES, C. R. Modern Processes in the Development of Pharmaceuticals. **Quimica Nova na Escola**, n. 3, p. 43-49, 2001.

RODRIGUES, C. R. F. et al. Quantum biochemistry study of the T3-785 tropocollagen triple-helical structure. **Chemical Physics Letters**, v. 559, p. 88-93, 2013.

ROSENBLOOM, J.; HARSH, M.; JIMENEZ, S. Hydroxyproline content determines the denaturation temperature of chick tendon collagen. **Archives of Biochemistry and Biophysiscs**, v. 158, n. 2, p. 478-484, 1973.

ROZAS, I.; ALKORTA, I.; ELGUERO, J. Bifurcated hydrogen bonds: three-centered interactions.

Journal of Physical Chemistry A, v. 102, n. 48, p. 9925-9932, 1988.

RUSSO, E. Chemistryplans a structuraloverhaul. **Nature Jobs**, v. 419, n. 6903, p. 4-7, 2002.

SANDHU, S. V. et al. Collagen in Health and Diseases. **Journal of Orofacial Sciences**, v. 2, n. 3, p. 153-159, 2012.

SANJEEV, J. et al. Collagen: Basis of Life. **Universal Research Journal of Dentistry**, v. 4, n. 1, p. 1-9, 2014.

SANTOS, H. F. The Concept of Molecular Modeling. **Cadernos Temàticos de Quimica Nova na Escola**, n. 4, 2001.

SANTAMARIA, R. et al. Molecular electrostatic potentials and Mulliken charge populations of DNA mini-sequences. **Chemical Physics**, v. 227, n. 3, p. 317-329, 1998.

SCHUMACHER, M. A.; MIZUNO, K.; BACHINGER, H. P. The crystal structure of the collagen-like polypeptide (glycyl-4®-hydroxyprolyl-4®-hydroxyprolyl)9 at 1.55 Å resolution shows up-puckering of the proline ring in the Xaa position. **Journal of Biological Chemistry**, v. 280, n. 21, p. 20397-20403, 2005.

SCHUMACHER, M. A.; MIZUNO, K.; BACHINGER, H. P. The crystal structure of a collagen-like polypeptide with 3(S)-Hydroxyproline residues in the Xaa position forms a standard 7/2 collagen triple helix. **Journal of Biological Chemistry**, v. 281, n. 37, p. 27566-27574, 2006.

SHOULDERS, M. D.; HODGES, J. A.; RAINES, R. T. Reciprocity of steric and stereoelectronic effects in the collagen triple helix. **Journal of the American Chemical Society**, v. 128, n. 25, p. 8112-8113, 2006.

SHOULDERS, M. D.; RAINES, R. T. Collagen structure and stability. **Annual Review of Biochemistry**, v. 78, p. 929-958, 2009.

SHOULDERS, M. D.; RAINES, R. T. Interstrand Dipole-Dipole Interactions Can Stabilize the Collagen Triple Helix. **Journal of Biological Chemistry**, v. 286, n. 26, p. 2290522912, 2011.

SILVA, T. H. A. Molecular modeling with the aid of computers. In: PRADO, M. A. F.; BARREIRO, E. J. (Org.). **Pràcticas de quimica farmacèutica y medicinal**. 1 ed. Pamplona: Cyted Sub-Programa X Quimica Farmacèutica, p. 67-85, 2002.

SPASSOV, V. Z.; YAN, L.; FLOOK, P. K. The dominant role of side-chain backbone interactions in structural realization of amino acid-code. ChiRotor: A side-chain prediction algorithm based on side-chain backbone interactions. **Protein Science**, v. 16, n. 3, p. 494506, 2007.

STOLLE, C. A. et al. Synthesis of an altered type III procollagen in a patient with type IV Ehlers-Danlos syndrome. A structural change in the alpha 1(III) chain which makes the protein more susceptible to proteinases. **Journal of Biological Chemistry**, v. 260, n. 3, p. 1937-1944, 1985.

SUPERTI-FURGA, A. et al. Ehlers-Danlos syndrome type IV: a multi-exon deletion in one of the two COL3A1 alleles affecting structure, stability, and processing of type III procollagen. **Journal of Biological Chemistry**, v. 263, n. 13, p. 6226-6232, 1988.

SUPERTI-FURGA, A. et al. Molecular defects of type III procollagen in Ehlers-Danlos syndrome type IV. **Human Genetics**, v. 82, n. 2, p. 104-108, 1989.

SUPERTI-FURGA, A.; STEINMANN, B. Impaired secretion of type III procollagen in Ehlers-Danlos syndrome type IV fibroblasts: correction of the defect by incubation at reduced temperature and demonstration of subtle alterations in the triple-helical region of the molecule. **Biochemical and Biophysical Research Communications**, v. 150, n. 1, p. 140-147, 1988.

SUZUKI, E.; FRASER, R. D. B.; MACRAE, T. P. Role of hydroxyproline in the stabilization of the collagen molecule via water molecules. **International Journal Biological Macromolecules**, v. 2, n. 1, p. 54-56, 1980.

TKATCHENKO, A.; SCHEFFLER, M. Accurate molecular van der waals interactions from ground-state electron density and free-atom reference data. **Physical Review Letters**, v. 102, n. 7, p. 073005, 2009.

TROMP, G. et al. A single base mutation that substitutes serine for glycine 790 of the alpha-1(III) chain of type III procollagen exposes an arginine and causes Ehlers-Danlos syndrome IV. **Journal of Biological Chemistry**, v. 264, n. 3, p. 1349-1352, 1989.

TROMP, G. et al. Sequencing of cDNA from 50 unrelated patients reveals that mutations in the triple-helical domain of type III procollagen are an infrequent cause of aortic aneurysms. **Journal of Clinical Investigation**, v. 91, n. 6, p. 2539-2545, 1993.

TROMP, G. et al. Substitution of valine for glycine 793 in type III procollagen in Ehlers- Danlos syndrome type IV. **Human Mutation**, v. 5, n. 2, p. 179-181, 1995.

UNIPROT CONSORTIUM et al. UniProt: a hub t for protein information. **Nucleic Acids Research**, p. gku989, 2014.

VESENTINI, S.; REDAELLI, A.; GAUTIERI, A. Nanomechanics of collagen microfibrils. **Muscles, ligaments and tendons journal**, v. 3, n. 1, p. 23, 2013.

VITAGLIANO, L. et al. Structural bases of collagen stabilization induced by proline hydroxylation. **Biopolymers**, v. 58, n. 5, p. 459-464, 2001.

WALTERS, B. D.; STEGEMANN, J. P. Strategies for directing the structure and function of three-dimensional collagen biomaterials across length scales. **Acta Biomaterialia**, v. 10, n. 4, p. 1488-1501, 2014.

WANG, H. L.; MACNEIL, R. L. Guided tissue regeneration. Absorbable barriers. **Dental Clinics of North America**, v. 42, n. 3, p. 505-522, 1998.

WANG, X. et al. The roles of knitted mesh reinforced collagen-chitosan hybrid scaffold in the one-step repair of full-thickness skin defects in rats. **Acta Biomaterialia**, vol. 9, n. 8, p. 7822-7832, 2013.

WAHL, M.C.; SUNDARALINGAM, M. C-H...O hydrogen bonding in biology. **Trends in Biochemical Sciences**, v. 22, n. 3, p. 97-102, 1997.

YAMADA, S. et al. Potency of Fish Collagen as a Scaffold for Regenerative Medicine. **BioMed Research International**, v. 2014, Article ID 302932, 2014.

YAN, R. B. et al. Rational design and synthesis of potent aminoglycoside antibiotics against resistant bacterial strains. **Bioorganic & Medicinal Chemistry**, v. 19, n. 1, p. 3040, 2011.

YANG, W. et al. Gly-Pro-Arg Confers Stability Similar to Gly-Pro-Hyp in the Collagen Triple-helix of Host-Guest Peptides. **Journal of Biological Chemistry**, v. 272, p. 2883728840, 1997.

YEH I. C.; LEE, M. S.; OLSON, M. A. Calculation of protein heat capacity from replicaexchange molecular dynamics simulations with different implicit solvent models. **Journal of Physical Chemistry B**, v. 112, n. 47, p. 15064-15073, 2008.

ZANATTA, G. et al. Antipsychotic haloperidol binding to the human dopamine D3 receptor: beyond docking through QM/MM refinement towards the design of improved schizophrenia medicines. **ACS Chemical Neuroscience**, v. 5, n. 10, p. 1041-1054, 2014.

ZHANG, D. W.; ZHANG, J, Z, H. Molecular fractionation with conjugate caps for full quantum mechanical calculation of protein-molecule interaction energy. **Journal of Chemical Physics**, v. 119, p. 3599-605, 2003.

ZHANG, D. W.; ZHANG, J. Z. H. Full quantum mechanical study of binding of HIV-1 protease drugs. **International Journal of Quantum Chemistry**, v. 103, p. 246-257, 2005.

ZHANG, J. et al. Protein folding simulations: from coarse-grained model to all-atom model. **IUBMB Life**, v. 61, n. 6, p. 627-643, 2009.

ZHOU, T.; HUANG, D.; CAFLISCH, A. Quantum mechanical methods for drug design. **Current Topics in Medicinal Chemistry**, v. 10, n. 1, p. 33-45, 2010.

Printed by Books on Demand GmbH, Norderstedt / Germany